P9-AQT-771

MANAGING THE CONSTRUCTION PROJECT

JOHN WILEY & SONS, INC.

CONSTRUCTION BUSINESS AND MANAGEMENT LIBRARY

COST ENGINEERING FOR EFFECTIVE PROJECT CONTROL
Sol A. Ward

MANAGING CONSTRUCTION CONTRACTS:
Operational Controls for Commercial Risks, Second Edition
Robert D. Gilbreath

MANAGING THE CONSTRUCTION PROJECT:
A Practical Guide for the Project Manager
Theodore J. Trauner, Jr., P.E., P.P.

MANAGING THE CONSTRUCTION PROJECT: A Practical Guide for the Project Manager

THEODORE J. TRAUNER, JR., P.E., P.P.

JOHN WILEY & SONS, INC.
New York / Chichester / Brisbane / Toronto / Singapore

In recognition of the importance of preserving what has been written, it is a policy of John Wiley & Sons, Inc., to have books of enduring value published in the United States printed on acid-free paper, and we exert our best efforts to that end.

This publication is designed to provide accurate and authoritative information in regard to the subject matter covered. It is sold with the understanding that the publisher is not engaged in rendering legal, accounting, or other professional services. If legal advice or other expert assistance is required, the services of a competent professional person should be sought. *From a Declaration of Principles jointly adopted by a Committee of the American Bar Association and a Committee of Publishers.*

Library of Congress Cataloging in Publication Data:

Trauner, Theodore J.
Managing the construction project : a practical guide for the project manager / Theodore J. Trauner, Jr.
p. cm. — (Construction business and management library)
Includes bibliographical references and index.
ISBN 0-471-55762-5 (cloth)
1. Construction industry—Management. I. Title. II. Series.
TH438.T728 1993 92-17788
690′.068—dc20 CIP

Printed in the United States of America

10 9 8 7 6 5 4 3 2

This book is dedicated to my daughters,
Stephanie, Melanie, Lorayne, and Michelle,
whose smiles make everything worthwhile.

CONTENTS

PREFACE

The focus of this book is the pragmatic application of management principles to the successful administration of a construction project. Most texts have addressed the topic of managing the construction project in a sterile theoretical fashion. This book offers a hands-on, common-sense approach to effective management. It is geared to the daily work that takes place on the project and presents suggestions and considerations that will facilitate the successful completion of the job.

Chapters 1 and 2 present the thought process the manager of construction will need in order to understand the roles of the parties to the project and how the various perspectives will affect the interaction and decision-making that will take place. Chapter 3 presents the kinds of approaches to management structure available, with an insight on how these may affect project execution, recognizing the biases the various parties bring to the project.

The "rules" for a construction project are embodied in the contract; Chapter 4 reviews some key considerations in the structuring of the contract. The essence of this discussion is to develop a contract that is practical and effective and not overly legalistic.

Chapters 5, 6, and 7 address the practical aspects of scheduling, managing changes, and preparing the appropriate level of documentation and records. Unlike most treatises that cover these areas with rules and lists, this discussion focuses on practical considerations that offer real rewards to the user.

Other potential problem areas are addressed in Chapter 8, including termination/default, claims, retainage, punchlists, and sole-source items. Again, the approach focuses on common-sense decision-making, which is the key to successful management.

Chapter 9 summarizes the key elements of the construction manager's role.

ACKNOWLEDGEMENTS

The author would like to acknowledge the following individuals for their capable assistance: Frances Murtaugh, who spent countless hours typing, editing, assembling materials, and coordinating this book; Donna Lane, who provided the graphics artistry, which makes this book easier to understand; and Dan Sayre and the editors at Wiley, whose objective feedback and professional editing were essential to the quality of the final product.

INTRODUCTION

This book describes how to manage a construction project. In the past, many such projects have been managed with a "seat of the pants" style. Contracts were handshakes. Schedules existed only in the manager's head. Budgets were gross in nature and cost monitoring was performed on a broad scale and in a haphazard manner.

Times have changed. The construction industry has been forced to improve its management abilities and to approach a construction project with far more skill and sophistication. The overall key to a successful project is thorough planning and precise execution. But in order to better plan and execute a project, the managers must be educated in areas that historically have been learned only through the school of hard knocks—on the construction site. Today's project manager must be well versed in contracts, the meaning and administration of changes, advanced scheduling techniques, and proper methods of information management. This book fills in the void of formal information dealing with these areas.

Since every construction project has multiple players with slightly different goals, Chapter 1 discusses the perspectives of the various players in terms of what success means to them. The owner, designer, and contractor-construction manager are included in this discussion. Finally, the chapter highlights the biases at play, depending on which party is handling the management of the project.

Chapter 2 emphasizes the forgotten fact that construction is a risky business. Therefore, management of the project is an exercise in risk management. The risks of each player are discussed, along with ideas on how to control and reduce risk. The important point that risk can never be eliminated is the focal point of this discussion.

On a broad scale, the various methods of managing a project are addressed so that both experienced managers and novices can gain a better perspective of approaches to managing a project. This leads to a detailed discussion of improving management through better contracts and better schedules. You will gain a clear understanding of the role that the contract plays in the process and how a practical contract should be structured to facilitate more efficient and successful management of the project. In the same vein, the importance of good scheduling is stressed since the schedule represents the detailed plan of execution for the project, both in terms of logical sequence and accomplishment of time objectives.

Since no construction project will ever be designed with perfect plans and specifications, changes are inevitable. The understanding, recognition, and proper management of change orders are vital to a successful project. All players must have a thorough understanding of this facet of construction in order to prevent a threat to the viability of the project.

Superimposed on all of the above areas is the requirement that a project be documented and effective systems be instituted such that key data is maintained in an orderly and efficient manner. The management of information must be planned for prior to the project and not just evolve as the project progresses.

Finally, this book identifies common problem areas that can occur on a project and suggests how to avoid them or resolve them with minimum impact to the success of the project.

The goal of this book is simple. It is to help the manager of a construction project make informed and intelligent decisions by providing ideas and insights to influence the way the manager views the project and formulates strategies to enhance success.

The role of a manager is simple. It is to make decisions. Those decisions control the course that the project will take. Those decisions determine the degree of risk to which the participants in the project are exposed. Those decisions ultimately mean the difference between success and failure.

CHAPTER 1

DEFINING SUCCESS

A high level of optimism always accompanies the start of a construction project. The taste of success is sweet and anticipation fuels an energetic beginning.

Each party involved with a project usually defines success differently. Thus, before embarking upon a discussion of how a construction project manager can improve the chances of a project's success, it is important to understand how success is perceived. Only after establishing a definition can a reasonable method be developed for achieving success.

Success or failure on a construction project depends on many factors. To a large extent it is a function of the perspective of the party viewing it. Thus, any discussion of success must focus on the perspectives of the various players on the construction project. By understanding not only one's own perspective of success but also those of the other parties, we can best measure our achievement during the course of the project.

The management team of most construction projects consists of three primary participants: the owner, with the need for the project; the designer, translating the owner's needs into tangible plans and specifications; and the contractor, making these plans and specifications a reality. The perspectives of each of these players vary, but an understanding of all is essential to any definition of success.

THE OWNER'S VIEW

The owner's perspective is based on need and on the capital and time available to meet that need. All three factors figure prominently in an owner's perception of success. The three factors are related, with variations in one impacting the other two. Need is first and foremost in the owner's mind. The need for the project is what initiated the entire construction process. Therefore, the ability of the project to fulfill the owner's needs is the owner's most important measure of success. Clearly, if the project does not function at an acceptable level, completion under budget or ahead of schedule is meaningless. However, it is rare that a constructed project fails to meet the functional needs established by the owner. Therefore, the focus gravitates primarily to costs/budget and secondarily to time.

Budget

The vast majority of projects are dollar driven. During the needs analysis for the project, the owner should have established a project budget. That budget may be based on many factors, but the bottom line that is established is consistently used as a measure of the project's success during the course of the job. The ever-present question is, "Will my project be built on budget, below budget, or beyond budget?" Of even more significance may be the follow-up questions concerning the ramifications of these three potential answers.

If the project is completed on budget, the owner should meet its plan and recognize the profits it anticipated. This would be particularly true if the owner bases the original budget on clearly and completely defined needs and profitable returns on investment. Bringing a project in on budget does not guarantee success. The two related factors of time and need must also be considered. An owner may complete a project within budget, but not meet the time goals that were set. As a result, though on budget, the owner may miss a window of opportunity and thus not realize the profitability that was planned. The author experienced this while consulting on a microconductor manufacturing facility. The budget was on track, but the project was behind schedule. As a consequence, the owner had to spend more in order to accelerate the project and secure a competitive position in the industry in terms of initial supply and sales of the product. This was clearly a case where being on budget alone would not be enough to achieve the desired goals.

Should the project be completed below budget, the cost savings should allow for an even greater margin of profit or larger return on investment. But again, this criteria cannot be taken alone without a consideration for time and need. While the possibility of greater returns are evident from completing a project below budget, there is no guarantee that this will occur. Therefore, the owner's expectations must be tempered by consideration of all three aspects of project management.

Of course, the owner's perceived worst case is to complete the project over budget. This introduces a new set of considerations and dilemmas. Perhaps the two most significant factors that the owner must consider are whether an amount over budget can be funded and at what point the project is no longer cost effective.

During a project's planning stage, an astute owner will consider the two items noted above: How much additional funding beyond prudently established contingencies is available and at what cost? Should the project experience overruns at any stage of the process, what point mandates that the owner halt the entire undertaking?

Time

As with the budget, the owner must consider the ramifications of deviations from the schedule. For example, what happens if the project is completed early? Most would respond that this is the best of all worlds and that if it ever happened to them, they would break out the party hats and noise makers. But such a reaction might be shortsighted. The owner must consider the consequences of an early finish. For instance, if the facility is completed early, can the owner utilize it at that time for its intended purpose. For example, if a manufacturing facility completes early but the plant equipment is not yet ready for installation, little or no advantage is gained by the early completion. In fact, the owner may have some additional cost since it will now be required to secure the new facility with no productive value being received. Similarly, an owner who has a new office building complete ahead of schedule may have to maintain two facilities at the same time if vacating the old facility was neither planned nor can be accomplished until a later date. Because early completion affects the budget, at least in the contingency area, and may also affect how the contract documents are structured at the project inception, it must be considered by the owner.

Of course, the most common case is that the project finishes later than planned. Numerous consequences can result from this. Obviously, the owner will continue to incur costs related to construction in such areas as the project designer's fee, financing costs, administrative costs, and even lost production or lost opportunity costs. As was noted earlier, finishing late may cause the project to miss a window of opportunity and push the return on investment below acceptable levels. In extreme cases, no return on investment may be realized.

Finally, a late finish to a project may result in the owner being exposed to requests for additional compensation from the contractor for delay damages. These can be significant and will be reflected in the return on investment. Time is as significant as budget. The owner must consider it at the planning stage and throughout construction. In essence, time is money.

Needs

An owner must properly assess the true needs of or functions for the project. This means that the owner must completely define the scope of the project—what it wants to build, when it must be built, and within what budget. Though this, too, sounds like a definition of the obvious, far too many owners do not consider "what happens if" early in the planning stage. A careful consideration of what, when, and how much will lead an owner to plan a project more completely and to have a more thorough understanding of the ramifications of deviations from plan.

In defining the functions of the project, the owner must consider all possible alternative courses of action to meet its needs or achieve its goals. This is not as simple as it may sound. For instance, an owner of a manufacturing facility decided to build a more efficient production line to process its raw materials. The project concept called for the construction of a new production line in the existing plant. The requirement was to design, fabricate, and deliver the new process equipment, shut down the existing production line, install the new equipment, and start up the refurbished plant. The existing plant could continue producing goods while the design, fabrication, and delivery of the new equipment was taking place. The project allowed the contractor 90 days to shut down the plant and change the equipment. The 90-day goal was not achieved. Instead, getting the new equipment on line required several more months. During that time the owner was unable to produce goods and suffered a loss in excess of $25,000,000. Did alternatives exist?

While many alternatives did exist, a rather obvious one was overlooked during the planning process. The new production line utilized some of the existing equipment. This was a factor in the decision to plan the project in the manner described above. As an alternative to tearing down the existing line and installing a new one, ample room was available in the plant to build the entire new line adjacent to the old one. By doing this, the owner's risks would have been negated. The old line could supply product while the new line was constructed, tested, and started. Since this approach would require some redundancy of specific pieces of equipment, the initial cost would be higher. An after-the-fact assessment of this approach showed that the redundancy would have increased the project budget by approximately $250,000. This was neither significant in the original project cost nor in comparison to the ultimate losses experienced. As obvious as this approach sounds, no one considered it during the planning stages for the project.

Goals

In simple terms, the owner must define the goal of the project and rigorously consider every approach that will achieve that goal. When those alternatives are defined, the owner can then perform a form of cost-benefit analysis, and then an informed decision can be made. That decision must allow for potential tradeoffs among budget, time, and need. Let's consider those for a moment.

In assessing the goals of the project, the owner may determine that the project must be completed faster than the traditional design-then-build schedule will permit. Consequently, the owner may decide to expedite the project. Gaining time by virtue of a fast-track project normally results in slightly higher construction costs. Presumably, this additional cost will be offset by some gain in reduced financing or more rapid generation of sales. The fast-track mode will occasion more changes during the construction project and will demand more up-front planning and greater management control during the course of the job. In turn, these considerations and decisions will dictate how the construction contract may be written and also how the owner will run the project in the first place. For instance, a fast-track approach may result in the owner utilizing a construction manager and issuing separate bid packages for phases of the work. This is significantly different from soliciting a single general contractor based on a completed set of plans and specifications.

Risk

The owner must assess what risk it is undertaking when beginning a construction project. That assessment of risk is effectively an evaluation of the chances for success. In the most general sense, the owner is focused on three overriding concerns:

1. The project will cost more than planned and budgeted.
2. The project will not be completed on time.
3. The project will not fulfill the defined needs or meet management's expectations.

The owner's perspective must be as broad as possible during the inception of the project. The more thorough the planning is, the better chances are for success.

THE CONTRACTOR'S VIEW

Like the owner, the contractor is also driven by considerations of dollars, time, and need. Though the elements are the same, the contractor's perspective of these is markedly different.

Cost

When a contractor reflects on the monetary aspect of a project, its first consideration concerns the contractual relationship that it is entering. The risks and hence the potential profits are markedly different for various types of construction contracts. Perhaps the best example of this is a comparison of *fixed-price*, or *lump-sum* bids with *cost-plus* contracts.

Fixed-Price Contract. When a contractor is bidding a fixed-price contract, many factors will affect the bidding strategy. First, it recognizes that it is agreeing to perform the work as specified for the stated lump sum. That amount can be altered only by changes to the contract, so the bid must be precise and accurate within the strategy adopted by the contractor. In formulating that bid, the contractor will evaluate several areas that will affect the price.

First, the contractor must determine how badly it wants the work. Obviously, the more desirable the project, the more the contractor will be willing to bid at a lower profit, no profit, or even at a loss. Prestige, new markets, and new areas of work are some of the factors the contractor must consider.

Prestige is normally associated with projects that have high visibility locally or even nationally. For instance, the project to refurbish the Statue of Liberty received national recognition. Contractors who participated in that project have used it widely in their advertising and promotion. Hence, their desire to win the project was high and could well have motivated low bids. Prestige can also be associated with prominent clients, such as the Department of Energy, Walt Disney Corporation, and NASA.

Often a contractor decides to branch out into new markets. A heavy and highway contractor may decide to create an industrial division and construct water treatment facilities. Since the contractor does not have a track record in that area, it may accept a lower profit in order to win work, establish a reputation, and broaden the company portfolio. Similar expansions are undertaken by firms that have been subcontractors but decide to bid work as general contractors instead. Similarly, a contractor may decide to expand its geographical base. For instance, a northeastern contractor may decide to expand into the sunbelt. Again, because of a lack of a track record and a base of operations, the bidding strategy may be influenced in order to win work in the new area. The entry into a new market affects the contractor's perspective regarding short-term dollars versus long-term returns of the expansion.

The considerations noted above are realistic, and low bids motivated by them are acceptable. A major problem occurs, however, if the contractor forgets the original bid strategy once the project has begun. If a contractor is bidding work low in order to achieve a long-term goal, then it must be remembered that the project will not now return a high profit. If it was bid originally as a loss, the actual occurrence of a loss should be expected and accepted.

The bid of a project is not only influenced by how much the contractor wants the work, but also by how badly it needs the work. During slow periods, a contractor's bid strategy may be affected by the need to keep its permanent staff employed. Even if a project is bid at no profit, a break-even situation may allow the company to maintain key staff for better times. Similarly, some projects may be bid as fill-in work between larger projects. These also allow full utilization of staff even if they don't generate profits.

There is one significant risk that a contractor is exposed to that is fostered by a bid strategy that takes work cheap. This is the tendency to win work in order to cover short-term cash-flow problems. In some situations, contractors have found that their cash flow was not sufficient to cover their operating costs. Though this should be a short-term occurrence, a contractor may bid work low in order to be able to generate cash for the corporation. The generation of cash can occur because contractors generally have some front-end loading on a project. This is a very dangerous situation. When a contractor bids a project low solely in order to generate cash flow, as the project progresses the cash flow will move into a deficit position. Since the projects were front-end loaded, the latter portions of the projects will require more cash than the respective jobs will generate. This can only exacerbate the overall negative cash position of the company. If the contractor has done this for too many projects, the ultimate outcome may be a financial disaster.

Presuming that the contractor does not have reasons such as those noted above for aggressively bidding work, other monetary factors must be considered. Of those factors, the one that should be at the top of the list is a consideration of changes on the job. In the most professional approach to the project, a contractor should insure that the construction contract includes an equitable pricing mechanism for changes. Anyone involved in construction recognizes that no projects are perfect and changes will occur. Even in the haste to win the job, the contractor must not ignore how those changes will be priced. Should the dollar value of the changes be significant, this may well affect the profitability of the project. Chapter 6 discusses changes in detail.

A philosophy has been expressed concerning changes and profits that bears some mention at this point. That philosophy proposes that the contractor should bid low on a project in order to win the job and then look to generate profits from changes. As many owners have expressed it, "the contractor lowballs the job and makes its money on the changes." This approach is suicidal. The contractor who bids low and intends to make its money on the changes is taking a big risk. First, there is the risk that there will be enough changes to generate the revenue to have a profit. Second, there is the risk that the pricing for the changes will be liberal enough to allow this to occur. Finally, there is the risk that the changes will be quickly resolved and paid in a timely fashion. In fact, most changes are resolved and paid long after the work is actually performed. Therefore, the contractor must finance those costs for that period of time. To approach a job with the belief that the profits will be made on changes demands that the contractor have a sound financial position and be prepared to collect its money well after the costs are expended. The author does not recommend this approach and this approach is not common practice in the industry.

The contractor must review the entire construction contract to assess how it affects the risk and chance of success and, consequently, its bid. A contract that is very one-sided in favor of the owner should cause the contractor to include more contingencies in the bid. Similarly, a contract that is rampant with exculpatory language might alert the contractor that the owner will be reluctant to pay any more than is absolutely necessary to complete the project. Hence, the contractor's bid may need to be higher. If a contractor is asked to assume all risks associated with a differing site condition for instance, then a prudent bid would incorporate some contingency for this. Likewise, if a contract includes a "no damage for delay" clause, the bids may need to be higher. The dollars of the job are a direct function of the language in the owner's construction contract.

Cost-Plus Contract. In the purest sense, in a cost-plus project the contractor is paid for the costs that are incurred in order to construct the project. The costs are defined in the contract and normally include the costs for materials, labor, equipment, the contractor's staff, and general conditions items. In practice, most cost-plus projects are not as simple as this. Often, cost-plus jobs are modified to be cost-plus fixed fee or cost-plus fixed fee with a guaranteed maximum price. The use of a fixed fee is intended to prevent the contractor from overstaffing the project and to encourage it to staff the project with the most efficient mix of personnel. Hence, the contractor's staffing costs, overhead, and profit are incorporated into the fixed fee. The owner's use of a guaranteed maximum price (GMP) is intended to place a cap on project costs. Obviously, using a fixed fee and a GMP will affect the ultimate bid that is submitted by the contractor.

The fixed-fee portion of the bid is subject to the same perspective noted in the discussion concerning the fixed-price contracts. Desire to get the work and/or the need to win the work, and similar considerations will be viewed in much the same light. The use of a GMP will influence the way contingencies are handled in the bid.

Another way to structure a bid is to use a "sharing in the savings" clause on a cost-plus fixed-fee contract with a GMP. In this, the contractor and the owner share in the Savings that are realized if the project is completed below the GMP that is established. The opportunity to share in the savings will affect the establishment of the GMP. Since a cost-plus approach is normally used for projects that are not completely defined in terms of final plans and specifications, the allowances that are used and the contingencies incorporated may be more liberal if savings can be shared. The amount of the potential share of the savings is measured against the risk of establishing the GMP in the first place. Also, the contractor's strategy on staffing and managing the project may be affected. Within this context, the contract documents again come into play. The contractor must determine whether the definitions for costs are equitable for all possible situations. Furthermore, the contract language concerning changes must be carefully reviewed to insure that changes are clearly defined with respect to the GMP. Oftentimes, a GMP includes allowances for items not fully designed. The contract language must differentiate between these items which are anticipated by the allowance and changes which may be initiated after the project is underway.

In bidding and estimating the specific items of work, the contractor attempts to be as accurate as possible for either the fixed-price or the cost-plus type of contract. The cost-plus contract may not have plans and specifications that are as completely defined as the fixed-priced contract. Therefore, the contractor is forced to utilize allowances and contingencies. In the fixed-price contract, the contractor will look for all possible methods to reduce its costs and, therefore, maximize profits. In a cost-plus job, the profits may be relatively fixed and, therefore, as long as costs are within the bid item amounts, there is little incentive to seek out cost savings. In both contract situations, the contractor still is bidding a job and any errors in that bid may result either in not winning the job or in losing money.

Time

Time is money for the contractor as well as for the owner. Hence, the contractor includes time as a key element of its perspective in approaching the construction project.

In most cases, the construction contractor is extremely optimistic about completing a project on time. Contractors rely too much on the implied warrantee of the plans and specifications. If a contract states that 350 calendar days are required to perform the work, the contractor believes that it can be performed in that amount of time. Only when the contractor begins to plan for an early completion is it likely to recognize exactly how much time the project will realistically take on a normal time basis. This occurs frequently because the contractor rarely lays out a tentative schedule prior to the submission of the bid. If it is bidding an early completion and hence reducing general conditions, it is more inclined to plan time more carefully and will then recognize any potential problems.

In a careful review of the contract requirements, the contractor should evaluate if the contract allows sufficient time to complete the project on a straight-time basis. If there appears to be a necessity for overtime or multiple shifts, it must factor this into the bid.

In general, the contractor focuses attention on evaluating the time required after the project is awarded, when it is required to submit a detailed schedule for the project.

There is one situation where the contractor does expend more effort in the consideration of time: when the bid requires both a specified cost and a duration. In other words, the contractor is bidding both time and money. Then the contractor must assess the relative tradeoff between time and cost. A faster schedule may result in some increase in the cost. In these situations, the contractor should attempt to determine how the owner will measure the relative weight of these two factors during the evaluation of bids.

The contractor's perspective of time is heightened if the contract documents contain clauses for liquidated damages or for incentives/disincentives. Usually, a "liquidated damage" clause does not give much cause for concern unless the stated amount is significant. In that case, the contractor may be motivated to carefully assess the time required for the project and may even include some contingency for liquidated damages within the bid. The use of an "incentive/disincentive" clause often sparks a more rigorous review of the time constraints. Usually the contractor will bid the project in anticipation of making the date necessary to achieve the stated incentive. For both of these types of clauses, the contractor, during the course of the project, will be more keenly aware of the schedule and will request time extensions should the situation warrant. In this manner, the contractor protects against a liquidated damage or a disincentive and attempts to maintain the chance of achieving the incentive.

Overall, the contractor's perspective of time is not as significant as the owner's at inception of the project. The importance of time grows as the project progresses, especially if any slippages in the schedule occur.

Needs

The reader will recall the earlier discussion of how much the contractor wanted the work and how much it needed the work. What the contractor must not lose sight of is that if it decides to fulfill needs over profits, a rational assessment must be made of the thoroughness and merit of the initial needs analysis.

The contractor must consider another element which relates to the needs of the project. This is the need as expressed in the contract's scope of work. Many construction projects utilize a complete design package and the contractor's obligation is to build the project in accordance with those plans and specifications. Some projects, however, will utilize performance specifications as opposed to design specifications. The performance approach delineates how the project or portions of the project must perform. It is up to the contractor both to design and construct those portions of the project that are described by the performance specification. For these situations, the contractor's consideration of needs is heightened. The contractor must insure that it understands exactly what is required and explore the alternative means of fulfilling those requirements. Once the alternatives have been defined, the contractor can evaluate them in terms of cost and time and reach a decision that can be factored into the bid. In essence, for performance type contract requirements, the contractor must fully understand the owner's needs, must define those needs in constructable terms, and then must evaluate the cost and time implications.

THE DESIGNER'S VIEW

The designer's view of success includes several elements, but at the top of the list must be the desire to do a good job. That means preparing a set of plans and specifications that fulfill the owner's needs in terms of function and cost. Within that desire to meet the needs is the motivation to make an idea become a reality, to take a concept from the owner and give it form and shape in a manner that allows it to become a definitive constructable project. While achieving all of this, the designer also wants to create something that is unique, will be respected, and will be representative of its design skills. This latter desire is oftentimes difficult to achieve because of the practical constraints of time and budget placed on the designer.

In striving for success, the designer must develop an accurate understanding of the owner's needs. Effectively, it must "crawl into the owner's head and look out through the owner's eyes." At that point, the designer can supply the creativity and technical skills the owner lacks and design the project.

The success of the designer, however, is not measured by the production of schematic drawings or even final design packages. The real "proof of the pudding" is determined at the completion of the project. It is then that the owner really "sees" the project and can assess the overall outcome in terms of cost and function. Thus, the designer must track the path to success throughout the entire construction process. During that process several other concerns are integral to the designer's measure of success.

Time

The first area that the designer must evaluate is the time available from the conception of the project to the date when design documents must be completed. This timeframe drives the decisions as to the project delivery system and the amount of resources the designer must devote to the project. Clearly, one stage of success is to have the contract documents prepared on time. Thus, time is tied to the designer's perspective of success. Also, the owner's budget for the design phase will influence the designer's ability to meet both the time constraints and the creative goals discussed above. Therefore, the three elements of need, time, and budget are present even in the first portion of the designer's equation for success.

These same three factors continue to influence the designer's ability to succeed throughout the remainder of the project. For instance, the bids must come in within budget or redesign efforts may be required. The project must be constructable within the project's overall time frame. And the incidence of change during construction must not be so high as to cause problems in budget and time. Obviously, the project as constructed must function as desired.

Since striving for success bespeaks a degree of risk, it is worthwhile to glance at the designer's perceptions of risk. The designer primarily is concerned with risk in terms of liability for design omissions and errors. The secondary focus is on the designer's profitability of the project.

Profit

To some extent, the perspective of the designer is shaped by the manner in which its services are contracted. Many design firms secure new work by some form of competitive process. Initial "cuts" or "short lists" are made based on demonstrated performance, qualifications, past experience, staff, and so on. The next phase of the process, however, is the determination of the design fees or method of compensation. In this phase, it is common for the owner to negotiate the best possible price for the design services. Effectively, the designer is being asked to take the work cheap.

In response to the negotiation process, the designer may well acquiesce to the owner's positions and indeed, take the work cheap. The designer rationalizes this position in several ways. First, the designer will view the work as filler work to be used by the staff when there are lulls in the more profitable billable hours. Next, the designer may decide that he can use the work as a training ground for his newer or junior staff. In some cases, the designer believes that it is worth the investment to take the work cheap because it may then have an advantage in getting future work from that owner. This is based on the philosophy that there is profit from volume. Unfortunately, volume work on a cheap basis still does not yield profits. Volume times zero is still zero.

Work undertaken for a low price normally is performed in a manner commensurate with the compensation. As a consequence, the project may not receive the attention of high-level staff nor the manhours for the most complete preparation of contract documents. The end result is that more changes, errors, and omissions occur. The designer then must be concerned about its liability for these.

Understanding the perspective of the designer is a function of knowing how a design firm realizes profits. Normally, design firms recognize profits as a direct function of the billable hours its staff can perform. The compensation is normally on an hourly basis in accordance with salaries and some defined multiplier. The multiplier has an element of profit built into it. Hence, the designer is motivated to keep its billable hours as high as possible.

Depending on the competition in the market, the designer may take on work that is beyond its normal expertise or perhaps beyond its staff level. The consequences of this are obvious. To perform the work, the designer must then bring on additional staff or subcontract the work. This may adversely affect the quality of the product.

Time is also of concern to the designer. The designer does not like to see a project take longer during the construction phase than anticipated since its schedule of work is usually established on a certain number of site visits, review of pay requests, and so on. If a project is delayed, the designer will be required to perform more of these tasks. It is often difficult to convince the owner that it should then receive additional compensation.

Risk

Let's return to the element of liability. The designer, particularly in today's construction environment, is very concerned about being sued for errors and omissions. This concern is clearly reflected in the wording of the modern American Institute of Architects (AIA) contract documents. These documents attempt to insulate the designer from any liability. For instance, the designer no longer "inspects" the project. Instead, the designer may make periodic "visits." Because of the concern over lawsuits, the designer attempts to take a hands-off approach to the actual construction phase of the job. The belief is that if the designer is not involved and doesn't make decisions or give direction, then it can't be held responsible for any problems that occur. Nothing could be further from the truth. Because of this concern over litigation, the designer's perspective is biased toward as little involvement as possible once the design is completed.

SUMMARY

By understanding the perspectives of the three major players on the construction project, one can gain an insight into how they interact and how a project can be successful. These perspectives shape the relative levels of risk that exist on the project for the players. Though there are many facets of risk on a job, the main areas can best be visualized by three risk triangles. These triangles also influence the success of the project. Therefore, they are also success/risk triangles. Figure 1-1 expresses the overall goals of the project—needs, time, and cost. These

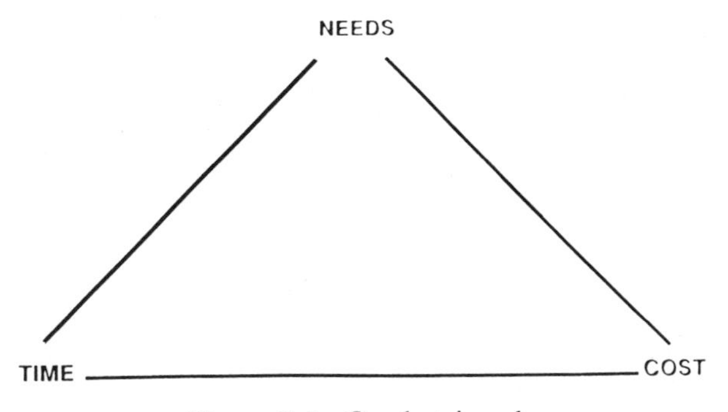

Figure 1-1. Goals triangle.

three goals must be realized for the project to be successful. Hence, any consideration of risk is understood within the confines of these three elements.

Figure 1-2 shows the interaction of the three players—the owner, the designer, and the contractor.

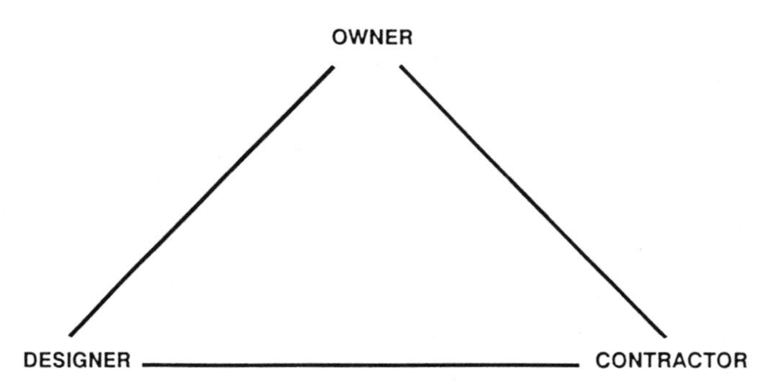

Figure 1-2. Players triangle.

Figure 1-3 portrays the major areas of the actual project that directly correlate with risk and success—the contract, the schedule, and the administration of change orders. It is within these elements that the players must operate.

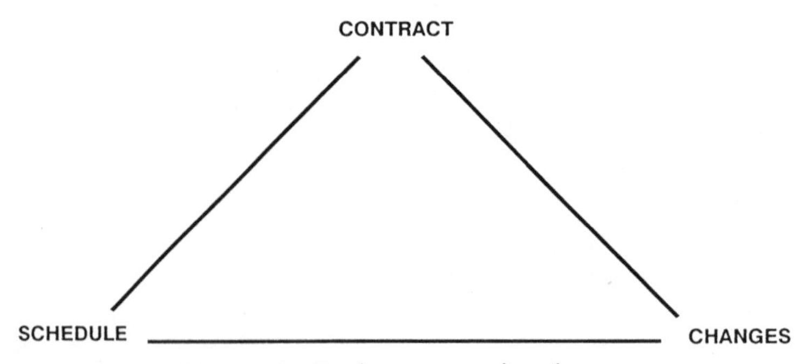

Figure 1-3. Project areas triangle.

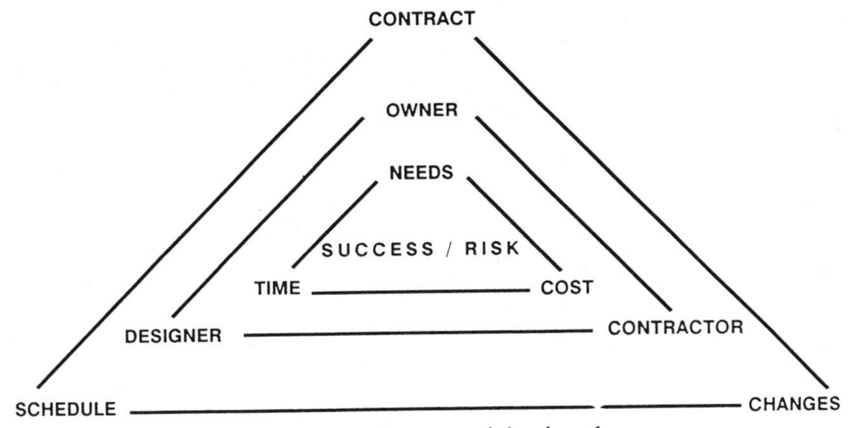

Figure 1-4. Success-risk triangle.

The combination of these three triangles, Figure 1-4, encapsulates the success/risk environment on the project. Risk is inherent on a construction project; it can never be eliminated. If risk is properly controlled and managed, however, success results.

CHAPTER 2

MANAGING RISK

All construction projects have an element of risk. The risk encompasses the elements of time, money, and quality, and the three are interrelated. The manner in which one manages and controls risk on a construction project is the subject of this chapter.

As an overview, the degree of risk is directly associated with the money spent and the time devoted to and the quality of the management of the project. The amount of each depends in turn on the manager's ability to identify the potential areas of risk in a timely fashion. The following discussion addresses two major areas that must be considered as part of a risk management program for a construction project: planning and the form of contract.

PLANNING

Planning must be done well up front to reduce the risk during the actual execution. The planning must be comprehensive and farsighted. Often, projects are best planned after a participant has had major problems on a preceding project because of a heightened sensitivity.

The owner must develop a "big picture" of the timing of the project in order to determine how it must pursue both the design and the construction. By establishing an overall schedule for the job, the owner can then determine the most appropriate project delivery approach. For instance, a project that has ample time from conception to delivery can proceed under a traditional design-then-build approach. Another project, however, that has a constrained timeframe for delivery may require a fast-track approach or phased construction. Similarly, another project may suggest that the owner pursue a design/build, or turnkey, approach. Perhaps it is wise at this stage to define these delivery methods since the terms have often been misused.

Traditional Design-Then-Build Approach

Historically, the traditional *design-then-build* approach has been the most prevalent form used for construction projects. In this method, the design is completed by a design firm before contractors are invited to bid on the work and before construction begins. Obviously, this requires ample time for a design to be developed and then time for the actual construction to occur. If the design requires a year and the construction two years, then the process would be as shown in Figure 2-1.

Fast-Track, or Phased Construction, Approach

Owners use the *fast-track*, or *phased construction*, approach in an attempt to have the project completed in a shorter timeframe than the traditional approach would require. For instance, in the example above, the project would take three years from the start of design to the completion of construction. This may not be acceptable to an owner in terms of its needs for the project. So, a method must be employed to decrease the duration for the job. In a fast-track job, the design and construction are integrated or overlapped so that the total time for the project is reduced. This is accomplished by breaking the project into specific phases and following the design of each phase with its con-

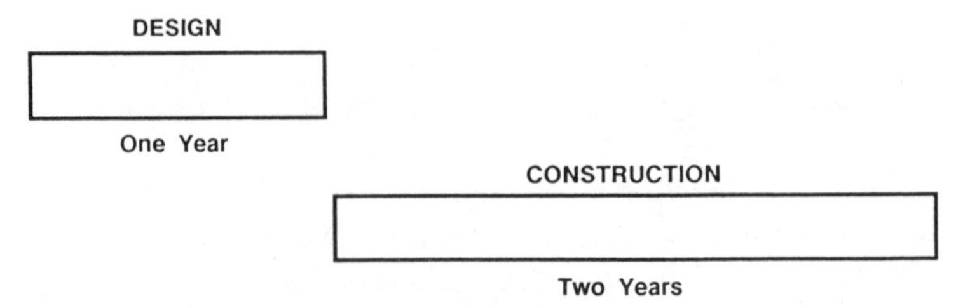

Figure 2-1. Design-then-build approach.

struction. While construction is occurring for the first phase, the design is being accomplished for the second. This process continues for the entire project. Taking the previous example, the project may be divided into the following phases:

1. Site work and foundations
2. Structural steel
3. Mechanical and electrical
4. Building enclosure
5. Interior finishes

The design would begin with the first phase, site work and foundations. As soon as that phase is complete, bids are solicited for construction. While the site work and foundations are being constructed, the design is being completed for the structural steel for the building. If properly timed, the structural steel package goes out for bid such that construction can begin approximately at the end of the construction of the foundations. The process continues as shown in Figure 2-2.

It is interesting to note that the term "fast-track" has been used many ways in the industry. Often, if a project is started and the drawings are not complete, it is called a fast-track project. This is not correct. Also, many people believe that the approach is relatively new, but it is not. The AIA has been using the term "phase construction" for many years.

It should be obvious that using a fast-track approach will result in a particular set of risks. Since the project's overall design is not complete when construction begins, there has been no opportunity to review and coordinate the drawings. As a consequence, there is normally a higher incidence of changes on such a project. Hence, the owner should enter the project with a larger contingency for the occurrence of changes.

Design/Build, or Turnkey, Approach

The third method of project delivery is a *design/build*, or *turnkey*, project. This approach awards a single contract for both the design and the construction of the project. By approaching the project in this manner, the contractor who will perform both functions can use a fast-track approach within its own organization in any manner that it sees fit, as long as it can accomplish the requirements of the contract. Usually, a design/build project is bid based on a set of performance specifications as opposed to design specifications. Performance specifications specify

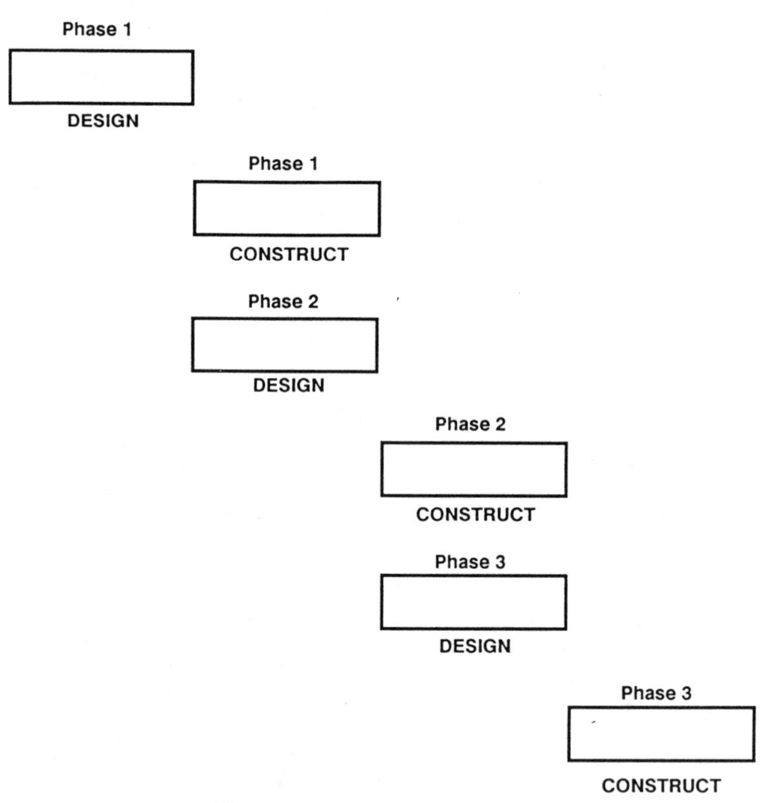

Figure 2-2. Fast-track approach.

the product that is desired and allow the contractor to design the system to meet those performance parameters. For instance, a performance specification may require a contractor to supply an HVAC system that meets certain temperature requirements for a specific facility. The contractor chooses how it will meet those performance requirements. Design specifications, on the other hand, spell out in detail exactly what has to be built. For instance, the specifications will state that the foundations will be spread footings with a certain thickness, dimensions, strength of concrete, and so on.

Historically, design/build projects have most often been used in the process industry. Recently, however, other types of projects have been utilizing a design/build approach. For instance, some Departments of Transportation have awarded design/build contracts for highway bridges. In essence, the DOT is drafting a specification that requires the design and construction of a bridge capable of carrying a certain number of vehicles, and so on.

Project Timing

It should be apparent now that by conceptualizing the overall project time requirements, the owner is directed toward a particular method of delivery. As we will see later on, this affects the type of contract that is written, the type of project management employed, and many other factors.

In developing an overall perspective of the timing of the job, it is beneficial to plan backward from the absolute "drop dead" date for when the facility must be available. By starting from the need date and inserting reasonable timeframes for the events that must occur, a decision maker may quickly realize that the traditional approach cannot be used.

SELECTION OF DESIGNERS

Presuming that one has successfully charted the time for the project, the work begins in earnest. At this stage, consideration must be given to selecting a designer for the job. This plays into the question of risk since a poor choice of designers may result in a total disaster. How can the owner reduce its risk in the process of employing a project designer?

Qualifications

The owner should consider some form of prequalification for the design firm. This prequalification can encompass many areas. At the minimum, the following areas should be addressed in the required submissions by the designers for them to prequalify:

Past work
Similar projects
References
Personnel
Workload
Financial capabilities

Most requests for credentials will require that the design firms submit detailed information on past projects on which they have worked. The requirement to submit this information should be specific so that the responses cannot be too general in nature. For instance, a firm responding to a request may list a large project that is germane to the one being planned. In reality, however, it may be that the firm was only a subconsultant for a small portion of the project cited. Consequently, the request should mandate that exact participation on any project cited be specific. In reviewing the information submitted, the projects should be viewed from the perspective of similar complexity, similar systems, etc. Many submissions will require that only similar projects be included in the information submitted. If this current project is a hospital, for example, then only medical facilities should be included in the proposal. This narrows the field and helps to avoid too much generalization.

For projects that are similar, the reviewer should verify that the competitive firms have actually performed the work and contact the project owners to determine the level of satisfaction concerning the designer on those jobs. In general, references should always be checked. Interestingly, the vast majority of competitive selection processes do not ever bother to check with the listed references.

It may be worthwhile for the reviewer to visit the offices of the firms that are in contention. During that visit, the number and type of personnel should be verified. It is also recommended that the turnover of personnel be checked. If it appears that the design firm has a high personnel turnover, continuity for your project and the quality of design staff may be a concern.

The present workload of the design firm may also be important. If the firm is extremely busy on a number of very large projects, perhaps your smaller project will not receive the management attention that you desire. A well-written request for proposal would require the design firm specify the personnel who will be assigned to the job. This allows the interviewer to meet with them during an office visit. The contract can also specify that personnel cannot be changed without the written approval of the owner.

In reviewing the staff of the design firm, an owner should verify that the key disciplines that are needed for the project are covered by the present staff. If the firm lacks mechanical or electrical designers, it should be apparent that the designer will be subcontracting out that portion of the work. If so, then these subconsultants should also be interviewed as part of the overall design team.

During the early planning stages, the owner should decide exactly what role the designer will play during both the design and the construction phase. This becomes particularly sensitive when the owner has some design or engineering capability on its staff. Clearly, the designer has the significant role during the design phase. The extent of involvement of the designer during the construction phase, however, is more difficult to decide. Chapter 3 discusses this in detail.

SCHEDULING THE DESIGN

Once a designer has been retained, there should be a clear, well-defined requirement for a schedule for the design. At best, most projects have a very sketchy schedule for the progress and completion of design. The schedule usually takes the form of specific dates for completion of various milestones, or phases, of design, for instance: schematics, 50% complete, 90% complete, and ready for bid. The problem with such a gross schedule is that there is virtually no way to review the status of the design and determine if the effort is on schedule or not. The schedule for design should entail as much effort as the schedule for construction. However the schedule is expressed—a detailed bar chart or a critical path method (CPM) schedule—it must have milestones defined so that one can review the work at hand and determine the status with respect to the schedule.

DESIGN INFLUENCE ON BUDGET

Earlier, we addressed the risks to the project in terms of cost. The owner is striving to complete the project within its budget. The budget must drive the design to a certain extent. Merely educating the designer as to the project needs, functions, and budget, may not be enough. It is not uncommon for the bids on a job to exceed what the owner has budgeted. This creates havoc as the owner and its designers must hustle to revise the design in an effort to value engineer the project to bring the initial cost estimate or bids within the allowable budget. An alternative should be considered to help prevent this from occurring.

The federal government utilizes a clause in its design contracts that mandates the design will be within the budget established. A sample of this type of clause is shown in Exhibit 2-1.

> The Design Contractor shall accomplish the design services required under this contract so as to permit the award of a contract, using standard Federal Acquisition Regulation procedures for the construction of the facilities designed at a price that does not exceed the estimated construction contract price as set forth below. When bids or proposals for the construction contract are received that exceed the estimated price, the Design Contractor shall perform such redesign and other services as are necessary to permit contract award within the funding limitation. These additional services shall be performed at no increase in the price of this contract. However, the Design Contractor shall not be required to per-

> form such additional services at no cost to the Government if the unfavorable bids or proposals are the result of conditions beyond its reasonable control.

Exhibit 2-1

Clearly, this form of clause puts a burden on the designer, but it is not an unfair burden. A prudent design firm will have a proven cost estimating capability to assess the value of its decisions during the course of the design. Given this concern for cost during the design stage, one question that should be asked of references given by a designer is how accurate its cost estimates were.

Designers also can affect the cost of the project in the area of changes. It is routine to have some changes occur during the course of construction. In fact, the owner budgets for this at the outset of the project. Therefore, one question that should be asked of references is what percentage of changes occurred on their projects with the specific design firm. The federal government has a policy of looking to the designer for the costs associated with errors and omissions. The Government guidelines are:

> Architect-Engineer Contractors shall be responsible for the professional quality, technical accuracy, and coordination of all services required under their contracts. A firm may be liable for Government costs resulting from errors or deficiencies in designs furnished under its contract. Therefore, when a modification to a construction contract is required because of an error or deficiency in the services provided under an architect-engineer contract, the contracting officer shall consider the extent to which the architect-engineer contractor may be reasonably liable. The contracting officer shall enforce the liability and collect the amount due, if the recoverable cost will exceed the administrative cost involved or is otherwise in the Government's interest.

Exhibit 2-2

The government also requires that a design contractor perform services to correct its errors at no cost to the government. Interestingly, on many occasions, an owner will pay for redesign services because of errors or deficiencies. This situation should be precluded by proper drafting of the contract with the designer.

In the process of contracting for design services, many owners desire to negotiate the most favorable price possible. There is nothing wrong with this philosophy, inherently; the owner is cautioned, however, not to be overly aggressive in this endeavor. The owner must recognize that most design firms estimate their costs based on the hours of effort that will be required to perform the design services. Coercing a design firm to lower its price may force the designer to reduce its hours. One of the easiest areas from which to eliminate hours is in the time allocated for coordination of design drawings. If this reduction is made in an attempt to cut cost, the natural result is that the drawings may not be as well reviewed and coordinated, thus causing more changes during the construction phase. The end result is that the owner will pay more rather than less, if the cost of the changes exceeds the original cost for more coordination efforts.

SITE STUDIES

During the planning stage, consideration must be given to the effort and money that will be devoted to studies of the site. Often, these investigations lack depth. This can cause major problems at the outset of construction. Unfortunately, many owners and designers try to keep costs down by skimping on items such as site investigations. The owner should consider what a good site study is worth. Let's look at an example.

Shortly after construction began on a construction project for an automated warehouse facility some unsuitable soils were discovered in the area where the building foundations were located. This necessitated a significant degree of rework, additional design effort, and caused a delay to the project of over six months. This should not have occurred.

Granted, a warehouse is not an extremely sophisticated project, nor does it normally merit a tremendous effort in terms of design or site studies. This warehouse, however, was designed to handle a computerized racking and stacking system. The robotics upon which the system were based required that the building slab had to be extremely level. Any deviation or settlement could adversely affect the functioning of the entire system. As a consequence of this, one would anticipate that the site investigations would focus particularly on the area of the slab. In this case they did not. In fact there were only three borings for the entire site and none of the borings were in the vicinity of the slab or its foundations.

The moral of the story is simple: site investigations are extremely important. This does not mean, however, that all projects require expensive presite work. Rather, the effort dedicated to the site studies should be a direct function of the nature of the site and the sensitivity of the design and the project function to potential site problems.

SELECTION OF CONSTRUCTION CONTRACTORS

Prequalifications should be followed for the construction contractors as well as for design contractors. This can be done not only for negotiated contracts, but also for competitively bid projects.

Many believe that a competitively bid project is open to all potential bidders and that no restrictions can exist as to who can participate in the bid process. This does not have to be the case. Should the owner decide, it can restrict the list of competitive bidders to those contractors who prequalify. On some occasions, even the federal government has followed this procedure. By first generating a list of qualified contractors, the owner eliminates the chance of having a low bidder who is totally out of its element.

Qualifications

Several areas should be explored during the prequalification stage. First is the obvious requirement to have all potential bidders provide information relative to past successful projects. This information should include such information as the size of the project, duration, date performed, owner, designer, and references. This information should be verified by the reviewer. In particular, references should be called to ascertain their perception of the performance of the contractor and to learn whether any problems occurred during the projects.

It is sometimes wise to visit other active projects that the contractor has ongoing. This allows the reviewer to see how the contractor organizes and manages the site, to view its work first hand, and generally to develop a feel for its on-site management.

Financial considerations should also come into play. The volume of past and present business should be provided in the initial information package. This allows the reviewer to assess where the project will fall into the overall corporate makeup. If the contractor has an annual volume of $200,000,000 and your project is a $2,000,000 job, there might be some concern as to the relative importance of your project to the overall organization. Similarly, if your project is worth $5,000,000 and the contractor's volume has been $2,000,000 per year, with its largest past project worth $950,000, you might question the contractor's ability to execute your larger project. This does not mean that you must choose a contractor who has done projects at least as large as yours. If this were a rule, no companies would ever grow. It is simply a point that should be considered, one which may provoke some additional investigations and questions.

The last area concerning finances would be an evaluation of whether or not the contractor has the financial ability to carry a project of your size. Commonly, there is some lag time between when a contractor performs work and is paid for it. The workers, subcontractors, and suppliers may require payment prior to this. Therefore, the contractor may have to carry some of the project costs for a limited period.

As with the designer, the contractor should specify the management staff who will be assigned to the project and make a written commitment that these people will remain on the project subject to the written acceptance by the owner of changes.

CONTROLLING AND MINIMIZING CHANGES

As noted in Chapter 1, changes are a major element of risk in a construction project. Therefore, in controlling risk, you must strive to keep the incidence of changes to a minimum.

Design Review

One cause of changes is errors and omissions in the plans and specifications. It must be recognized that no set of plans and specifications will be perfect, but they should be as error-free as possible. A consideration in this area is to have an outside or independent design review performed before the plans and specifications are issued for bid. This can even be an ongoing process while the design is progressing as opposed to a review when the design is complete. The review should note any ambiguities or conflicts detected. It is often beneficial if the reviewing firm looks at the documents as a contractor would when bidding and constructing the project. An outside design review probably will not detect all errors, but generally will more than pay for itself by uncovering a number of problems that can be resolved up front.

Constructability Review

Similar to a design review is a constructability review. This review of the project prior to bid insures that the project can be built as designed. That may sound facetious, but there have been far too many examples of construction projects that could not be constructed as designed. Besides the obvious constructability, the plans are reviewed to see if there are more cost-effective alternatives to the project. For instance, might a steel structure be less expensive than concrete or might prefabricated systems be employed that would lower cost or speed up construction time. During this process, one is effectively performing a value engineering of the job.

The review should also consider life-cycle costing. For instance, the initial cost of an HVAC system might be higher, but the maintenance cost might be lower and the life span longer. So, over the useful life of the project, the net costs are lower. Whether the owner decides to incorporate suggestions such as these is an economic decision. However, at the minimum, these alternatives should be considered.

PROJECT CONTRACTS

The contract plays a significant role in the risk management of the project. Hence, carefully drafted contract documents, particularly the contract General Conditions, are a major factor in controlling risk. These should be considered the "rules of the game." If a problem arises on the project, the General Conditions dictate how it will be resolved. Because of this, the General Conditions should be drafted with an eye to "what happens if." This exercise should consider delays to the project, default by the contractor, termination by the owner, reasons for time extensions, disputes and how they are to be resolved, and management and pricing of changes. Because of the significance of the contract to controlling risk, Chapter 4 discusses the contract in detail.

BONDS

Any discussion of the subject of risk must address bonds. Required by the owner, bonds on a construction project usually take three possible forms: bid bonds, payment bonds, and performance bonds. Bonding companies will not issue bonds to contractors unless there is a reasonable degree of certainty that the contractor will perform. They usually investigate the contractor carefully to insure that the bond is secure. Therefore, the owner benefits from the diligence of the bonding company in checking that the contractor has the capabilities to perform the work. Requiring bonds, however, increases the cost of the project by the amount of the bond.

The *bid bond* is a form of guarantee by contractors that assures the owner that the low bidder will accept award of the contract. Should the low bidder refuse to perform prior to award, the owner can seek recovery of the cost differential between the low bid and the second bid from the bonding company that has underwritten the bond.

Given the function of a bid bond, does this necessarily mean that it is always advisable to effect it if the low bidder is reluctant to perform? No, not always. Consider this scenario: an owner receives bids for a project from several contractors. The low bidder is so low that the owner questions whether a mistake might have been made or if there was some misunderstanding of the contract. The owner and designer meet with the low bidder and it is discovered that a mistake occurred or a misunderstanding existed. The owner may insist that the contractor perform or the bid bond will have to make up the difference between the low bid and the next lowest bid.

In response to this quandary, the contractor may elect to perform, knowing that it is going into a losing project. In this case, the obvious motivation for the contractor is to do whatever it can to make up the loss while the project is being constructed. This may lead to cutting corners or looking for as many changes or extras as possible. In consequence, the owner may end up with far more headaches than it anticipated. A more prudent approach might be to allow the contractor to withdraw its bid with no enforcement of the bid bond. This approach should only be allowed in rare circumstances, otherwise there would be no reason to have a bid bond in the first place. This approach may seem liberal, but it is one that is used by the federal government, which even allows upward correction of bids if the contractor can show the exact mistake that was made.

A *payment bond* is a form of insurance against the contractor not paying subcontractors, materialmen, or suppliers. Should the contractor not pay, rather than lien the project, the supplier can instead seek recovery against the payment bond.

Performance bonds are used as a form of guarantee on the actual performance of the project. The common belief is that if a contractor defaults on a construction project, the bonding company or Surety will complete the work. Though this most often occurs, it is not exactly what a performance bond states. Figure 2-3 is a sample performance bond for a project.

PERFORMANCE BOND *(See Instructions on reverse)*	DATE BOND EXECUTED *(Must be same or later than date of contract)*			
PRINCIPAL *(Legal name and business address)*	TYPE OF ORGANIZATION *("X" one)* ☐ INDIVIDUAL ☐ PARTNERSHIP ☐ JOINT VENTURE ☐ CORPORATION			
	STATE OF INCORPORATION			
SURETY(IES) *(Name(s) and business address(es))*	PENAL SUM OF BOND			
	MILLION(S)	THOUSAND(S)	HUNDRED(S)	CENTS
	CONTRACT DATE	CONTRACT NO.		

OBLIGATION:

We, the Principal and Surety(ies), are firmly bound to the United States of America (hereinafter called the Government) in the above penal sum. For payment of the penal sum, we bind ourselves, our heirs, executors, administrators, and successors, jointly and severally. However, where the Sureties are corporations acting as co-sureties, we, the Sureties, bind ourselves in such sum "jointly and severally" as well as "severally" only for the purpose of allowing a joint action or actions against any or all of us. For all other purposes, each Surety binds itself, jointly and severally with the Principal, for the payment of the sum shown opposite the name of the Surety. If no limit of liability is indicated, the limit of liability is the full amount of the penal sum.

CONDITIONS:

The Principal has entered into the contract identified above.

THEREFORE:

The above obligation is void if the Principal –

(a)(1) Performs and fulfills all the undertakings, covenants, terms, conditions, and agreements of the contract during the original term of the contract and any extensions thereof that are granted by the Government, with or without notice to the Surety(ies), and during the life of any guaranty required under the contract, and (2) perform and fulfills all the undertakings, covenants, terms conditions, and agreements of any and all duly authorized modifications of the contract that hereafter are made. Notice of those modifications to the Surety(ies) are waived.

(b) Pays to the Government the full amount of the taxes imposed by the Government, if the said contract is subject to the Miller Act, (40 U.S.C. 270a-270e), which are collected, deducted, or withheld from wages paid by the Principal in carrying out the construction contract with respect to which this bond is furnished.

WITNESS:

The Principal and Surety(ies) executed this performance bond and affixed their seals on the above date.

PRINCIPAL			
Signature(s)	1. *(Seal)*	2. *(Seal)*	*Corporate Seal*
Name(s) & Title(s) *(Typed)*	1.	2.	

INDIVIDUAL SURETY(IES)		
Signature(s)	1. *(Seal)*	2. *(Seal)*
Name(s) *(Typed)*	1.	2.

CORPORATE SURETY(IES)					
SURETY A	Name & Address		STATE OF INC.	LIABILITY LIMIT $	*Corporate Seal*
	Signature(s)	1.	2.		
	Name(s) & Title(s) *(Typed)*	1.	2.		

NSN 7540-01-152-8060
PREVIOUS EDITION USABLE

25-106

STANDARD FORM 25 (REV. 10-83)
Prescribed by GSA
FAR (48 CFR 53.228 (b))

Figure 2-3. A performance bond.

CORPORATE SURETY(IES) *(Continued)*						
SURETY B	Name & Address			STATE OF INC.	LIABILITY LIMIT $	*Corporate Seal*
	Signature(s)	1.	2.			
	Name(s) & Title(s) *(Typed)*	1.	2.			
SURETY C	Name & Address			STATE OF INC.	LIABILITY LIMIT $	*Corporate Seal*
	Signature(s)	1.	2.			
	Name(s) & Title(s) *(Typed)*	1.	2.			
SURETY D	Name & Address			STATE OF INC.	LIABILITY LIMIT $	*Corporate Seal*
	Signature(s)	1.	2.			
	Name(s) & Title(s) *(Typed)*	1.	2.			
SURETY E	Name & Address			STATE OF INC.	LIABILITY LIMIT $	*Corporate Seal*
	Signature(s)	1.	2.			
	Name(s) & Title(s) *(Typed)*	1.	2.			
SURETY F	Name & Address			STATE OF INC.	LIABILITY LIMIT $	*Corporate Seal*
	Signature(s)	1.	2.			
	Name(s) & Title(s) *(Typed)*	1.	2.			
SURETY G	Name & Address			STATE OF INC.	LIABILITY LIMIT $	*Corporate Seal*
	Signature(s)	1.	2.			
	Name(s) & Title(s) *(Typed)*	1.	2.			

BOND PREMIUM ▶	RATE PER THOUSAND	TOTAL
	$	$

INSTRUCTIONS

1. This form is authorized for use in connection with Government contracts. Any deviation from this form will require the written approval of the Administrator of General Services.

2. Insert the full legal name and business address of the Principal in the space designated "Principal" on the face of the form. An authorization person shall sign the bond. Any person signing in a representative capacity (e.g., an attorney-in-fact) must furnish evidence of authority if that representative is not a member of the firm, partnership, or joint venture, or an officer of the corporation involved.

3. (a) Corporations executing the bond as sureties must appear on the Department of the Treasury's list of approved sureties and must act within the limitation listed therein. Where more than one corporate surety is involved, their names and addresses shall appear in the spaces (Surety A, Surety B, etc.) headed "CORPORATE SURETY(IES)". In the space designated "SURETY(IES)" on the face of the form insert only the letter identification of the sureties.

(b) Where individual sureties are involved, two or more responsible persons shall execute the bond. A completed Affidavit of Individual Surety (Standard Form 28), for each individual surety, shall accompany the bond. The Government may require these sureties to furnish additional substantiating information concerning their financial capability.

4. Corporations executing the bond shall affix their corporate seals. Individuals shall execute the bond opposite the word "Corporate Seal"; and shall affix an adhesive seal if executed in Maine, New Hampshire, or any other jurisdiction requiring adhesive seals.

5. Type the name and title of each person signing this bond in the space provided.

☆ U.S.G.P.O.: 1987 - 181-247/60143

STANDARD FORM 25 BACK (REV. 10-83)

Figure 2-3. *(Continued)*

In essence, a performance bond is a guarantee of performance in the penal amount of the bond. If the contractor defaults, there is no assurance that the Surety will complete the work, only the assurance that it will be liable for the amount stated in the bond.

CONTRACT FORMS

The form of contract used for a project can also influence the degree of risk. While there are many variables in how a contract can be structured, the most common types are addressed below.

Lump-Sum Contracts

The most common form of construction contract is a lump-sum, or fixed-price, contract. In this type of contract, the contractor submits a lump-sum amount to perform the work required in the contract documents. Clearly, the burden of risk is on the contractor to perform within the lump sum it bid. To use this form of contract, the owner must have a complete design before soliciting bids. Because of this, a lump-sum contract may not be feasible for a fast-track project.

Cost-Plus Contracts

In a cost-plus contract, the owner pays the contractor the costs that the contractor incurs in accomplishing the work, plus a fee or percentage for managing the work and for profit and there is no specified limit to the contract amount. In this instance, the burden of risk has shifted to the owner.

In order to put some limits on what the cost of a project will be under a cost-plus contract, many projects use a cost plus a fixed fee with a GMP. In this type of contract, the contractor is paid for its costs and is also paid a fee for the management of the project, and performs the work within the GMP that is agreed to.

Cost-plus and cost-plus with GMP contracts are usually used when the design is not complete, the project is a fast-track job, or for design/build projects. In these cases, the contractor may be dealing with more conceptual or performance type documents and is basing the GMP on experience with similar systems and projects. In this case, the risk is shared both by the contractor and the owner.

It must be noted that there is a misconception concerning the meaning of a GMP. A GMP does not mean that the price is guaranteed and no matter what it will never go any higher. It means that the price is limited to the GMP amount if the owner makes no changes to the project. Any changes incurred will increase the GMP amount, just as they would for a lump sum contract. Some problems can occur because of this and the parties must be cognizant of them when preparing the contract.

For many cost-plus GMP contracts, a "sharing in the savings" clause may be incorporated. An example of this type of clause is shown below.

> *Contract Savings:*
>
> Should the costs for the completed project as invoiced by the construction manager (CM) in accordance with clause 8.1 Cost of the Work be less than the contract GMP, then the owner and CM will share in the savings. The distribution of all savings less than the GMP will be 60% to the owner and 40% to the CM. The owner will pay the CM any savings due after final project auditing is complete.

This kind of clause notes that if the contractor can complete the work for a cost below the GMP, any savings will be shared between the owner and the contractor in the specified ratio. This is intended to create an incentive for the contractor to bring the project in as economically as possible. It may, however, also create a situation where the contractor may seek increases to the GMP through the changes mechanism. By doing this the contractor is attempting to maintain a cushion in order to collect as much of a bonus as possible through the "sharing in the savings" clause.

How is a GMP bid if the contract documents aren't complete? Normally, the contractor includes some form of contingency in its bid for the items that are not fully detailed. For instance, the HVAC or electrical systems may not be designed. Based on similar projects, the contractor estimates what a reasonable cost will be for the undesigned systems. A problem may occur when the design is completed. The contractor may assert that its contingency or allowance envisioned a less costly approach, and the system designed thus is a change, and should increase the GMP.

Consider the example of a hospital project in the midwest. The contract was a cost-plus with a GMP. It included a sharing in the savings clause but limited the maximum amount the contractor could realize to $250,000. To maximize its return, the contractor used every opportunity to increase the GMP by describing items as changes. In reality, many of the items were not changes but should have been included under the specified allowances and contingencies.

Many owners believe that any form of cost-plus contract is undesirable. However, if the nature of the project warrants it, a cost-plus approach may be the only reasonable method to use. For instance, a project for the design and construction of a process plant was performed on a cost-plus basis since the process was a pilot and had never been done before. As a consequence, it was recognized that many unknowns were involved. Therefore, the owner could either deal with continuous changes as the project evolved or adopt a cost-plus basis. Another project involved the removal of hazardous wastes from a site. Because of time constraints, it was not possible to define all types of waste that would be encountered or the quantities involved. Also, the owner wanted the work carefully supervised by an environmental consultant. It was further recognized that the remediation methods to be utilized for the hazardous wastes might vary depending on the materials encountered. For these reasons, a cost-plus approach was prudent.

Obviously, a degree of caution must be exercised to embark on this more open-ended method of construction. For example, in an area where casino gambling had been newly legalized, almost frantic efforts were undertaken to build the new casinos and begin operation. One owner began a project on a cost-plus basis since virtually no plans had been developed for the casino. As the project progressed, the plans were about finalized when construction was approximately 50% complete. At this point, the construction costs were high and the owner stopped all work and required the contractor to supply a lump sum bid to complete the remaining work. It took about eight weeks for the contractor to provide the estimate. This apparent long period was occasioned by the fact that the contractor had to determine exactly what was in place as compared to the plans, define the work remaining, and then estimate it. Once the bid was submitted, the owner recognized that it did not have the financial ability to finish the project and work stopped indefinitely.

Before an owner decides to pursue construction on a cost-plus basis, careful thought should be given to the possible risks versus the presumed benefits. The contract documents must clearly define what will be considered costs. Though this sounds simple, it is not. The examples below illustrate this.

Cost of the Work:

11.4 The term Cost of the Work means the sum of all costs necessarily incurred and paid by CONTRACTOR in the proper performance of the Work. Except as otherwise may be agreed to in writing by OWNER, such costs shall be in amounts no higher than those prevailing in the locality of the Project, shall include one the following items, and shall not include any of the costs itemized in paragraph 11.5.

11.4.1 Payroll costs for employees in the direct employ of CONTRACTOR in the performance of the Work under schedules of job classifications agreed upon by OWNER and CONTRACTOR. Payroll costs for employees not employed full time on the Work shall be apportioned on the basis of their time spent on the Work. Payroll costs shall include, but not be limited to, salaries and wages plus the cost of the fringe benefits which shall include social security contributions, unemployment, excise and payroll taxes, worker's compensation, health and retirement benefits, bonuses, sick leave, vacation and holiday pay applicable thereto. Such employees shall include superintendents and foreman at the site. The expenses of performing Work after regular working hours, on Sunday, or legal holidays, shall be included in the above to the extent authorized by the OWNER.

11.4.2 Cost of all materials and equipment furnished and incorporated in the Work, including costs of transportation and storage thereof, and manufacturers' field services required in connection therewith. All cash discounts shall accrue to CONTRACTOR unless OWNER deposits funds with CONTRACTOR with which to make payments, in which case the cash discounts shall accrue to OWNER. All trade discounts, rebates and funds, and all returns from sale of surplus materials and equipment shall accrue to Owner and Contractor shall make provisions so that they may be obtained.

11.4.3 Payments made by CONTRACTOR to the Subcontractor for Work performed by Subcontractors. If required by OWNER, CONTRACTOR shall obtain competitive bids from Subcontractors acceptable to CONTRACTOR and shall deliver such bids to OWNER who will then determine, with the advice of ENGINEER, which bids will be accepted. If a subcontract provides that the Subcontractor is to be paid on the basis of Cost of the Work Plus a Fee, the Subcontractor's Cost of the Work shall be determined in the same manner as CONTRACTOR'S Cost of the Work. All subcontracts shall be subject to the other provisions of the Contract Documents insofar as applicable.

11.4.4 Costs of special consultants (including, but not limited to, engineer, architects, testing laboratories, surveyors, lawyers, and accountants) employed for services specifically related to the Work.

11.4.5 Supplemental costs including the following:

11.4.5.1 The proportion of necessary transportation, travel, and subsistence expenses of CONTRACTOR'S employees incurred in discharge of duties connected with the Work.

11.4.5.2 Cost, including transportation and maintenance of all materials, supplies, equipment, machinery, appliances, office, and temporary facilities at the site and hand tools not owned by the workmen, which are consumed in the performance of the Work, and cost less market value of such items used but not consumed which remain the property of CONTRACTOR.

11.4.5.3 Rentals of all construction equipment and machinery and the parts thereof whether rented from CONTRACTOR or others in accordance with rental agreements approved by OWNER with the advise of ENGINEER, and the costs of transportation, loading, unloading, installation, dismantling, and removal thereof—all in accordance with terms of said rental agreements. The rental of any such equipment, machinery, or parts shall cease when the use thereof is no longer necessary for the Work.

11.4.5.4 Sales, use, or similar taxes related to the Work, and for which CONTRACTOR is liable, imposed by any governmental authority.

11.4.5.5 Deposits lost for causes other than CONTRACTOR'S negligence, royalty payments, and fees for permits and licenses.

11.4.5.6 Losses and damages (and related expenses), not compensated by insurance or otherwise, to the Work or otherwise sustained by CONTRACTOR in connection with the execution of the Work, provided they have resulted from causes other than the negligence of CONTRACTOR, any Subcontractor, or anyone directly or indirectly employed by any of them or for whose acts any of them may be liable. Such losses shall include settlements made with the written consent and approval of OWNER. No such losses, damages, and expenses shall be included in the Cost of the Work for the pulse of determining Contractor's Fee. If, however, any such loss or damage requires reconstruction and CONTRACTOR is placed in charge thereof, CONTRACTOR shall be paid for services a fee proportionate to that stated in paragraph 11.6.2.

11.4.5.7 The cost of utilities, fuel, and sanitary facilities at the site.

11.4.5.8 Minor expenses such as telegrams, long distance telephone calls, telephone service at the site, expressage, and similar petty cash items in connection with the Work.

11.5.6 Other overhead or general expense costs of any kind and the costs of any item not specifically and expressly included in paragraph 11.4.

11.4.5.9 Cost of premiums for additional Bonds and insurance required because of changes in the Work.

11.5 The term Cost of the Work shall not include any of the following:

11.5.1 Payroll costs and other compensation of CONTRACTOR'S officers, executives, principals (or partnerships and sole proprietorships), general managers, engineer, architects, estimators, lawyers, auditors, accountants, purchasing, and contracting agents, expediters, timekeepers, clerks, and other personnel employed by CONTRACTOR whether at the site or in his principal or a branch office for general administration of the Work and not specifically included in the agreed upon schedule of job classifications referred to in subparagraph 11.4.1—all of which are to be considered administrative costs covered by the CONTRACTOR'S Fee.

11.5.2 Expenses of CONTRACTOR'S principal and branch offices other than CONTRACTOR'S office at the site.

11.5.3 Any part of CONTRACTOR'S capital expenses, including interest on CONTRACTOR'S capital employed for the Work and charges against CONTRACTOR for delinquent payments.

11.5.4 Cost of premiums for all Bonds and for all insurance whether or not CONTRACTOR is required by the Contract Documents to purchase and maintain the same (except for additional Bonds and insurance required because of changes in the Work).

11.5.5 Costs due to the negligence of CONTRACTOR, any Subcontractor, or anyone directly or indirectly employed by any of them or for whose acts any of them may be liable, including but not limited to, the correction of defective Work, disposal of materials or equipment wrongly supplied, and making good any damage to property.

Contractor's Fee:

11.6 The Contractor's Fee allowed to CONTRACTOR for overhead and profit shall be determined as follows:

11.6.1 a mutually acceptable fixed fee; or if none can be agreed upon,

11.6.2 a fee based on the following percentages of the various portions of the Cost of Work:

11.6.2.1 for costs incurred under paragraph 11.4.3, the Contractor's Fee shall be five percent; and if a subcontractor is on the basis of Cost of the Work Plus a Fee, the maximum allowable to the Subcontractor as a fee for overhead and profit shall be ten percent, and

11.6.2.2 no fee shall be payable on the basis of costs itemized under paragraphs 11.4.4, 11.4.5, and 11.5.

11.7 The amount of credit to be allowed by CONTRACTOR to OWNER for any such change which results in a net decrease in cost, will be the amount of the actual net decrease. When both additional and credits are involved in any one change, the combined overhead and profits shall be figures on the basis of the net increase if any.

Exhibit 2-4

The clause states that the contractor is paid direct labor rates and a markup of 36% which includes 16% for payroll burdens. The payroll burden percentage was developed from such burdens as FICA, State Unemployment Insurance (SUI), and worker's compensation. These were all based on standard percentages. The actual costs, however, may not be the same. For instance, FICA is approximately 7.5%. It is limited, however, to a maximum amount paid depending on the total earnings of the individual for the year. It is possible and even likely that individuals will reach their maximum earnings subject to FICA before the end of the year. When this occurs, the 7.5% becomes a profit to the contractor. The same is true for SUI. SUI has a limit in most states that is significantly below the FICA limit. Therefore, when an individual reaches the maximum earnings subject to SUI, the percentage used in the calculation also becomes a profit for the contractor. In a similar manner, workers compensation also may have hidden profits. Normally the percentage used for workers compensation is calculated from published guidelines for various crafts. The actual charge to a contractor, however, is based on an experience factor which directly correlates to the number of claims made against the worker's compensation policy held by the company. The actual amounts paid may be as low as 50% of that calculated.

A more accurate approach to cost is to pay for the actual costs incurred, which can be substantiated by accounting records.

In short, costs may not be what they seem. Therefore, careful thought must be given to the drafting of the contract if a cost-plus form of contract is to be used. The contract must clearly define which costs are allowed and which ones are not. The definition of costs must be practical and must be auditable and trackable. In essence, if you draft a cost-plus contract, you should feel comfortable that every dollar spent can be identified and allocated to a specific definition of cost.

BIDS FOR COST AND TIME

One final area needs to be addressed concerning controlling risk and the contract form. This relates to projects where the owner is soliciting bids both for cost and for project duration. In these, the contractor submits a price for the work and specifies the duration in which the work can be performed.

If an owner is utilizing this approach, it must establish a basis upon which different bids can be compared. Without this, there is no way that the owner can make an accurate decision as to the best alternative to select. This means that the owner must develop a means to measure both time and dollars on a common scale. Usually, that common scale is dollars. In other words, the owner must be able to assign a dollar value to every day that the project is shorter than opposing bids. How this is accomplished depends primarily on the nature of the project. A manufacturing facility might be able to quantify the daily revenues earned and use that as a measure. An educational facility, however, might use a different dollar scale for time based on a window of opportunity to use the facility for various semesters during the year.

If time is an element of a bid, the owner may consider requiring the submission of a detailed schedule with the bid. This will allow the owner to decide if the contractor has reasonably planned the work such that it can be accomplished in the time bid. Unfortunately, this creates additional work and cost for the contractor during the bid process. It may even discourage some potential bidders from submitting a bid. An alternative is to conduct the bid process in two phases. In phase 1, the owner requires a bid based on time and dollars with either no schedule or only a summary schedule required. Once a preliminary evaluation and "first cut" has been made, a short list of contractors results. In phase 2, the short list of contractors is required to demonstrate the feasibility of their respective plans to complete the project within their timeframes and amounts bid.

SUMMARY

The concept of risk will continue to be a common thread throughout this book. It is managed and controlled by careful attention to many areas throughout the entire project. The success of a risk management program, however, is most significant at the beginning of the job. The initial planning and the development of the proper project delivery and resulting form of contract will set the stage for the continuing management of the project.

CHAPTER 3

METHODS OF MANAGING THE PROJECT

There are many ways in which a construction project can be managed. This chapter addresses the most common approaches of the project designer also providing construction services (the traditional approach), the general contractor managing the project, and a construction manager supervising the project. Any of these methods can and have been successful on projects. The propriety, benefits, and shortcomings of each of these will be noted in this chapter. The key element of this discussion is to develop the ability to recognize the potential biases and problems that may exist with each form of management. Conceptually, a party to a project can develop biases because of two elements: dollars and liability. We will refer to these two factors throughout the discussion.

THE TRADITIONAL APPROACH—MANAGEMENT BY DESIGNER

Historically, the majority of construction projects have been managed by the firm who performed the design services for the project. This is looked upon as a logical extension of its services since it was involved with the project almost from its inception. Normally, this work has been referred to as the project's *construction services* phase.

The involvement of the designer during the actual construction of the project can vary widely. It may range from periodic visits, perhaps only monthly, to having a full-time resident engineer and related support staff on site for the duration of the job.

There are several duties that the design firm normally will fulfill. It will review and approve the contractor's requests for payment. In this process, the designer is certifying that the amount requested is reasonable, that the work represented by the invoice has been performed, and that the performance of the work is in accordance with the contract requirements.

The designer will observe (as opposed to inspect) the work while it is in progress. This is done during the site visits. Not too long ago, design firms provided inspection services as part of the construction services phase. Unfortunately, because of litigation and liability concerns by designers, the function has more recently been described as observation. The intent is to avoid any liability if problems arise in the quality of the work performed. While this subtle difference has not been tested in the courts, at least by the writing of this book, it is likely that the designer will still retain the liability for problems that may occur if it has observed them but has not acted in accordance with industry standards to prevent or resolve the problem.

During the construction services phase, the designer will routinely initiate, review, estimate, and process any change orders or contract modifications that may occur. Likewise, the designer will conduct job site meetings and maintain and issue minutes from these meetings. The meetings function to monitor progress and problems and facilitate resolution of conflicts or questions that may arise. Finally, the designer will maintain all job files until they are turned over to the owner at the completion of the job.

Potential Designer Bias

The use of the designer during the construction phase of the project is an acceptable approach to the management of the project, and it has been used for an extensive period of time. Both the owner and the contractor should recognize, however, the potential biases that the designer may have during construction and the consequent risks associated with those biases.

During the construction phase, the designer should not be directly influenced by the dollars portion of the equation since it has already been paid for the majority of its design services and is most likely billing on an hourly basis or cost-plus basis for the field support provided. An exception to this would stem from the clauses in the design contract. If the design contract is written such that the designer is liable for the costs of errors or deficiencies, then a mixed dollar and liability bias may develop. Otherwise, the designer is most concerned about its liability during the construction phase. If this were not the case, projects would still be inspected and not just "observed."

The designer's potential bias to protect the integrity of its design may foster actions on the part of the designer that can create problems for both the owner and the contractor. For instance, should the contractor point out a problem that it believes is an error or an omission, the designer might respond that "A knowledgeable contractor should have known that it was the *intent* of the contract documents to include that item." Or, "You should have discovered that at the time of bid and raised the question then." Most construction contractors have experienced responses of this type. The reason the designer might respond in this fashion is that the designer is reluctant to admit that there may be a design problem, lest it be held liable for any increased costs. As a consequence of this bias, the contractor has the risk that an item that should be a routine change order and be paid for may be denied. This forces the contractor to absorb the costs or to assert a claim in accordance with the disputes procedures in the contract. Since the owner holds the contract with the contractor, it then has the risk of becoming embroiled in a dispute.

It is likely that many contractors and owners have observed that design firms during the construction services phase have a tendency to group changes together such that only a few contract modifications are issued. Though the designer may state that this saves time and paperwork, a more subtle reason exists. Consider for a moment that you are the owner of a construction project and your designer is managing the construction phase of the project, including the handling of change orders. Over the course of the job, 25 change orders or contract modifications are issued. Since the job was 2 years long, and based on your past experience, this does not seem to be an unreasonable amount. If, however, each one of those change orders was in reality a compilation of ten smaller changes, had they been issued separately you would have had 250 change orders on the project, and your reaction to this would be very different. Even though the dollar value is the same, with that volume of changes, you might question the ability of the designer and the credibility of the design itself.

For all of the reasons noted above, the designer in the role of manager during construction will want to have as few changes as possible. This does not necessarily protect the owner, as it can lead to a far more adversarial project or even to claims and disputes.

The existence of this bias does not mean that the designer should not be used during the management of construction. Rather, it requires that several factors be assessed before the designer is chosen as the manager.

Assessing Designer Bias

The owner must first evaluate the designer's staff that is to be assigned to the construction phase and, specifically, its field experience and level of commitment to the project. Oversight by an individual who has only design experience and virtually no field experience can lead to problems and an even greater manifestation of the bias noted above. This is amplified if the field staff was also involved in the actual project design.

Second, the owner must evaluate the designer's knowledge and facility in the area of construction scheduling and estimating. These two elements will be a daily concern during the active project. If the designer is without practical experience in these areas, the project may well end up with no viable method for monitoring progress and increase the prospect of arguments occurring over costs for changes. Often, the lack of expertise in the area of scheduling results in a poorly drafted scheduling requirement in the construction contract, which is then aggravated by inadequate methods of measuring progress. For example, many projects require that monthly status reports be prepared by the field representative. The status report often notes percentages of time elapsed versus money spent. "The project is behind schedule, since 87% of the contract time has elapsed but only 70% of the project has been completed." This 70% figure is based on approved pay requests, but this can be misleading because it presumes a linear relationship between time and money. Most often no linear relationships exists. Chapter 5 discusses this in detail.

Another common scheduling problem occurs when a contractor requests a time extension because of a change order or because extra work was needed. If the owner's representative responds that "no time extension is warranted because there was other work being performed during the period of the alleged delay," it bespeaks a lack of understanding of scheduling and, in particular, the critical path of a project.

To estimate costs, many design firm field representatives rely solely on estimating texts or manuals. Without the benefit of actual field experience, this can lead to problems. For example, a project that involved the rehabilitation and repair of a major highway bridge was being performed. After most of the project was completed, the designer made a change that added pipe from the bridge scuppers, or drains, down the piers. The pipe would act as a drainage system and prevent storm water runoff from falling on buildings below the bridge. The designer prepared an estimate based solely on a simple PVC pipe installation, but failed to consider how the system would be installed at that stage of the job. The new bridge deck was already in place, so there was no easy method to get below the bridge to install the pipe. The designer's estimate was approximately $14,000. The actual cost was almost $70,000. Had the estimate been closer to actual cost, an alternative approach might have been considered.

Third, the owner should evaluate its own ability to review and measure the effectiveness of the designer during the construction phase. Recognizing the potential biases, the owner must be prepared to monitor the project to identify if these biases are occuring. Many owners have no capability to do this; it is why the owner hires a manager in the first place. The owner might consider utilizing an independent agent for this review.

Fourth, the owner should assess the degree of responsibility the designer is willing to assume during the construction phase. It is common for a construction services agreement to be drafted by the design firm and to be rampant with disclaimers. You will see such clauses as:

> While the architect will conduct periodic observations of the project, this in no way guarantees the quality of the work or the method or manner of performance by the contractor. The contractor remains solely liable for full compliance with the contract requirements and the quality of the work performed.

Some design firms offer both design and construction services but separate the two functions within their organization. The selling point to the owner is that by having two separate divisions, the construction services group will be independent and will not be influenced by the design group. The relative truth of this position can only be assessed by the owner based on past performance of the firm. Reasonably, however, it is still the same firm with the same corporate profit motivation.

The owner must exercise care with respect to how the designer is compensated for the construction services phase. For example, in a project in which the designer agrees to perform construction services, its overall design contract might be based on a 7½% fee. In other words, the designer would receive 7½% of the *total* cost of the project, including any changes or extra work performed. Compensation on a percentage basis of this sort may encourage higher costs, because the higher the costs, the greater the fee to the designer.

MANAGEMENT BY THE GENERAL CONTRACTOR

In a perfect world, there would be no need for any form of project management other than that done by the general contractor performing the construction work. This would be so because in our perfect world we would have a perfect design, perfect contract documents, a complete project concept such that there would be no owner-initiated changes, and a perfect contractor performing perfect work. All the owner would have to do would be award the contract, pick up the keys to the facility at the project's completion, and, of course, pay the bills. Obviously, we do not live in a perfect world.

In all cases, the general contractor or the respective prime contractors are managing their portion of the construction project. In terms of the overall project, that management becomes even more significant if the designer makes only periodic visits and the owner has limited involvement.

Total management by the general contractor is really a matter of trust. Projects where an owner and contractor have an established relationship can utilize this kind of management. Generally, this can occur on repetitive projects where few if any changes are anticipated and the contractor has a track record for very similar work. A good example of this might be construction of chain stores or franchises such as fast food facilities, retail stores, and theaters.

Design/build projects are generally managed by the contractor since it is responsible for both the design and construction and they occur simultaneously. In most cases, however, turnkey jobs usually have a degree of complexity, and the owner has some form of representation on the project.

In essence, the use of the construction contractor to manage the job depends on the complexity of the project, the adequacy of the design, the standard for quality of the work, and the reputation and track record of the contractor. The owner's risk is a direct function of the professionalism and trustworthiness of the general contractor.

CONSTRUCTION MANAGEMENT

The term "construction management" is relatively new and given to numerous different interpretations. To address this area, we will first deal with the concept and the objectives, then the various forms of execution, and finally with the biases that may exist.

In the evolution of construction management, the original concept was that the CM would act as an extension of the owner—function almost as if he or she were the owner's staff—and have sole loyalty to the owner.

The CM was to be a firm with construction experience who would become involved with a project prior to contract award, preferably during the design phase, and even as early as the conceptual phase. In that manner, the CM could act as an advisor to th owner during the development of the project and follow the project through to its completion. By having the CM involved early on, the owner could benefit from a design review and a constructability review by the CM. Similarly, the CM could provide ideas during the feasibility study phase, value engineering and life-cycle costing during the design phase, assistance in contracting for the construction of the project, and management during the actual construction.

From an application standpoint, the professional construction manager, or agency construction manager, relationship has the following characteristics:

1. The CM works solely for the owner.
2. The CM does not do the design.
3. The CM does not contract or perform any of the actual construction work.
   A. The owner enters into a general contract for the construction work.
   B. Alternatively, the owner enters into prime contracts for the construction work.
4. The CM is truly a manager and not just a broker for prime contractors or subcontractors.

In practice, construction management has come to mean many different things. For instance, some design firms offer construction management services for the projects that they design. In reality, this is no different than the traditional approach, or construction services phase, that has been offered for many years. The same problems and biases prevail.

As CM became more popular, many construction contractors began offering CM services. The key to this form of CM is an understanding of the services that are offered and the contractual manner in which the relationship is set up. Let's explore some of the variations and note the potential problems or shortcomings.

The first variation has the CM entering into a lump-sum contract with the owner for construction of the project. Since the CM is performing the construction or at least subcontracting out all of the work, this relationship is really no different from having a standard general contractor on board.

The second variation is to have a CM awarded the construction contract on a cost plus a fixed-fee basis with a guaranteed maximum price, which is also no different from having a general contractor on board. Since the CM is still contracting to perform the construction, it will have the same biases as a general contractor. These will be discussed shortly. It should be recalled that the use of a cost plus a fixed-fee contract is dependent on the status of the contract documents and the nature of the project.

Potential CM Bias

Other than the professional CM approach described above, the nature of the services that the owner receives are largely dependent on how the CM is compensated for the work. For example, if the CM takes a contract on a cost plus a percentage basis, then there is bias toward wanting the cost to be higher. For every dollar added to the project, the CM receives increased compensation.

To counteract this bias, the owner may require a GMP, perhaps with a sharing in the savings clause. In this case, the CM may have a bias toward protecting the GMP by noting anything that remotely resembles a change in scope in order to increase the GMP amount. For instance, in a project involving the construction of new hospital facilities, the owner hired a CM. The contract stated that if the project was completed at a cost less than the GMP, the CM would receive the first $250,000 of the savings and 20% of any additional Savings. The contract documents were not fully detailed and, as a consequence, the GMP had numerous allowances and contingencies. As the project was constructed and the design parameters became better defined, the CM consistently requested change orders, contending that the specific design requirements were not contemplated within the GMP bid. Upon careful review, it was determined that many of these "changes" should have been included more appropriately in the allowances and contingencies. The CM was trying to protect the amount of savings.

There is an even more subtle bias that exists when the owner uses a company that offers both CM and general contractor services. If, in the CM role, the staff is compensated on some form of cost-plus basis, then that project may well be an acceptable place to utilize the second-level staff. Since that company may have a lump-sum project down the street and the owner's cost-plus job here, it is only natural that the best staff will be assigned to the lump-sum job. Clearly, this is not what the owner desires.

In essence, the best alternative for a construction manager's role is the professional CM approach previously described. There are no apparent biases in that relationship, and the owner should receive the representation and loyalty that it needs.

Before leaving the concept of CM, one further point is worthy of note. The Construction Management Association of America has used the phrase "owner advocacy" to describe the role that a CM should play, but the term "advocacy" is somewhat misleading. Normally this term would connote someone who will support and fight for the owner—no matter what. This is not what the owner should desire. The professional CM has an obligation to point out to the owner any problems that the owner may cause. An advocate will not usually do this. In other words, the construction manager should be an extension of the owner's staff but it should remain independent so that it can also be the owner's conscience.

PROGRAM AND PROJECT MANAGEMENT

Some construction projects have a project manager, or program manager. This is a matter of semantics. Often, the role played by a program manager (PM) is the same as that of a professional CM. In some rare cases, a program manager may play an even broader role than a CM would. For instance, a large school district faced with a significant expansion program hired a program manager to assist the district in solving its problems of meeting the rising demand. As such, the program manager was required to evaluate needs, capabilities, and propose short- and long-term solutions. The PM sought out qualified design firms for specific projects and oversaw the construction operations of the various facilities. If a distinction exists between CM and PM services, it is primarily in the breadth of activities and responsibilities.

THE OWNER AS MANAGER

There is no reason why an owner cannot manage the construction project without the assistance of outside help. In order to successfully accomplish this, however, the owner must have an experienced in-house staff. Needless to say, this means that the owner must also have a significant volume of construction work in order to make this approach cost effective.

If the owner chooses to run the project, it should first resort to a critical introspection to identify any shortcomings that might impair the efficient management of the job. At a minimum, the owner must assess time, and staff available and the level of experience within the organization.

It is advisable for owners to hire only those resources that are necessary. Those resources are determined first by assessing the needs of the project and comparing those with the existing capabilities of the owner's organization. The difference between the two is what should be hired. Too often, an owner will hire either more or less help than is really needed. Both situations result in extra cost to the owner.

Finally, the owner must recognize that if it chooses to manage the project itself, its staff on the job may be reluctant to recognize, accept, and pass on bad news. As a consequence, there is an inherent self-deception that particularly affects evaluations and decisions on budget and time. For example, in a project for a large pharmaceutical company, construction was falling further behind schedule as each month went by. To counter this, the owner's manager held the end date of the project static and either shortened activity durations or rescheduled such that more activities occurred concurrently. Suffice it to say, the project was late and over budget.

SUMMARY

What's the bottom line of this discussion? Forget all the buzz words and acronyms. Step back and assess what the project demands in terms of management capabilities. Evaluate what capabilities are on hand and seek assistance to bridge the gap between the two. In acquiring that assistance, structure a relationship such that any potential biases are minimized. Those biases that cannot be eliminated must be monitored to preclude them affecting the outcome of the job.

The evaluation of the method of managing the project should be related to your assessment of risk. If the choice of management approach is the traditional one, the risk of the designer's biases affecting changes and cost must be consciously considered. If the owner chooses to manage the project itself, the risks associated with lack of expertise should be weighed. Should the CM approach be used, the exact role and contractual relationship will similarly establish different risks.

As long as the owner understands the risks involved with the respective project delivery systems, he can prepare for them and exercise appropriate management controls to reduce them.

CHAPTER 4

THE CONTRACT AS A RISK MANAGEMENT TOOL

One of the main elements of the risk triangle described in Chapter 1 was the construction contract. The contract spells out the rules of the game. Therefore, it deserves a significant amount of attention if risk is to be reduced for all parties. It should be noted that reducing risk in construction contracts is not accomplished solely in the legal sense but also in the practical sense of effectively managing the project. This pragmatic approach is the focus of this chapter. It is best to begin with a basic discussion of construction contracts.

BASIC CONCEPTS OF CONSTRUCTION CONTRACTS

In its simplest form, a contract is an agreement between two parties. Normally it is a written document but can be binding even as an oral agreement. In general, when we agree to certain items in a contract, one party cannot unilaterally change this. For instance, if we sign a contract to buy a new car and the specified color is red, the automobile dealership cannot deliver a green car without violating the contract. You, as the buyer, do not have to accept the car. Interestingly, construction contracts do not follow this same general rule. In this respect, they are unique.

The main reason that construction contracts are unique is by virtue of the "changes" clause which is contained in virtually all of them. The changes clause allows the owner to make changes within the scope of the contract. The contractor will be compensated for the changes, but it must perform them or be considered in default. There is no requirement that the owner first seek the contractor's agreement as to the change. It should be noted, however, that the contractor is entitled to an equitable adjustment for the costs and time associated with the change. This unique feature, plus several other aspects, of construction contracts are why many projects end in dispute.

Construction contracts have another feature that helps foster adversarial relationships. That feature is the special role that the drafter of the contract plays. In most cases, the design contractor drafts the construction contract. In many situations, the contractor has no ability to negotiate the terms of the contract, particularly in public projects which are competitively bid. Thus, it may be somewhat one-sided from the outset. The designer generally fulfills the role of interpreter of the contract, and this is specifically stated in many instances. If a question arises as to the meaning of the contract, the designer renders a decision.

Finally, the designer is usually specified as the decision maker for appeals to any problems that arise. Examples of such clauses are shown below.

> 4.2.11 The Architect will interpret and decide matters concerning performance under and requirements of the Contract Documents on written request of either the Owner or Contractor. The Architect's response to such requests will be made with reasonable promptness and within any time limits agreed upon. If no agreement is made concerning the time within which interpretations required of the Architect shall be furnished in compliance with this Paragraph 4.2, then delay shall not be recognized on account of failure by the Architect to furnish such interpretations until 15 days after written request is made for them.
>
> 4.2.12 Interpretations and decisions of the architect will be consistent with the intent of and reasonably inferable from the Contract documents and will be in writing or in the form of drawings. When making such interpretations and decisions, the Architect will endeavor to secure faithful performance by both Owner and Contractor, will not show partiality to either, and will not be liable for results or interpretations or decisions so rendered in good faith
>
> 4.3.2 Decision of Architect. Claims, including those alleging an error or omission by the Architect, shall be referred initially to the Architect for action as provided in Paragraph 4.4. A decision by the Architect, as provided in Subparagraph 4.4.4, shall be required as a condition precedent to arbitration or litigation of a Claim between the Contractor and Owner

as to all such matters arising prior to the date final payment is due, regardless of (1) whether such matters relate to execution and progress of the Work or (2) the extent to which the Work has been completed. The decision by the Architect in response to a Claim shall not be a condition precedent to arbitration or litigation in the event (1) the position of Architect is vacant, (2) the Architect has not received evidence or has failed to render a decision within agreed time limits, (3) the Architect has failed to take action required under Subparagraph 4.4.4 within 30 days after the Claim is made, (4) 45 days have passed after the Claim has been referred to the Architect, or (5) the Claim relates to a mechanic's lien.

Exhibit 4-1

Given the role that the designer plays, the contract may be even more biased than noted in Chapter 3. Thus, this is an additional cause for adversarial relationships to develop.

Finally, the contract is a legal document. Therefore, when problems arise they are often settled in the legal system, which, by its very nature, is adversarial.

The goal of the successful project manager is to avoid conflicts and use the contract to control risk and reduce problems. This can be done and is worth the effort that it entails. The following discussion presents ideas and methods to point the reader in the right direction. Bear in mind that each project is unique and, therefore, your contract documents must be structured to meet the characteristics of your project. In all cases, it is strongly recommended that your contract be reviewed by a qualified construction attorney before you enter into it.

STANDARD CONTRACTS

There are several standard contract forms in use for construction projects. These include the American Institute of Architects (AIA) documents, the Associated General Contractors (AGC) documents, the Construction Management Association of America (CMAA) documents, and documents produced by professional engineering organizations such as the Engineers Joint Contract Documents Committee (EJCDC).

These standard contract forms are intended as a guideline for the formulation for a specific contract for construction. They are not intended to be used exactly as drafted. In fact, the standard documents do have some shortcomings, which the parties will want to avoid. It should also be noted that the standard contract documents reflect the biases of the organization that has promulgated them. Thus, the AIA documents are structured to protect the designer and not the contractor or the owner. Similar analogies can be drawn concerning the other sources.

While standard contract forms are an adequate starting point, they should not be used without being modified to suit your specific needs.

COMPOSITION OF THE CONTRACT

The contract is normally composed of the following parts:

An Agreement
The General Conditions
Special Provisions or Supplemental Provisions
Technical Provisions or Specifications
The Plans or Drawings

The Agreement normally is a short document executed by both parties that states the basic tenets of what is being agreed to. It references the other portions of the contract.

The General Conditions spell out the procedures for administration of the contract. This is the most important part of the contract and is the focus of most of this chapter.

The Special or Supplemental Provisions are used to supplement, modify, or expand the General Conditions. They are used to tailor a standard contract to a specific project or to add items which might not have been included in initial drafts of the contract.

The Technical Provisions, normally referred to as the specifications, describe in detail the technical portions of the work to be performed.

The plans or drawings give a pictorial representation of the project to be constructed.

SIGNIFICANT CONTRACT CLAUSES

While this book cannot address all the subtleties or nuances of a construction contract, the most significant considerations will be presented. Should the reader desire additional information, there are a number of texts that directly address the subject of construction contracts. The following clauses and variants of them will be addressed:

Changes
Differing Site Conditions
Scheduling
Time Extensions
Disputes
Liquidated Damages
Termination
Exculpatory Language

Changes

The changes clause in a construction contract may be the most important clause. This clause makes the construction contract unique from other forms of contracts. A typical one allows the owner to make changes in the scope of the work. An example of this type of clause is shown in Exhibit 4-2.

Article 10—Changes in the Work

10.1. Without invalidating the Agreement, OWNER may at any time or from time to time, order additions, deletions, or revisions in the Work; these will be authorized by Change Orders. Upon receipt of a Change Order, CONTRACTOR shall proceed with the Work involved. All such Work shall be executed under the applicable conditions of the Contract Documents. If any Change Order causes an increase or decrease in the Contract Price or an extension or shortening of the Contract Time, an equitable adjustment will be made as provided in Article 11 or Article 12 on the basis of a claim made by either party.

10.2 ENGINEER may authorize minor changes in the Work not involving an adjustment in the Contract Price or the Contract Time, which are consistent with the overall intent of the Contract Documents. These may be accomplished by a Field Order and shall be binding on OWNER, and also on CONTRACTOR who shall perform the change promptly. If CONTRACTOR believes that a Field Order justifies an increase in the Contract Price or Contract Time, CONTRACTOR may make a claim therefore as provided in Article 11 or Article 12.

10.3 Additional Work performed without authorization of a Change Order will not entitle CONTRACTOR to an increase in the Contract Price or an extension of the Contract time, except in the case of an emergency as provided in paragraph 6.22 and except as provided in paragraph 10.2 and 13.9.

10.4 OWNER shall execute appropriate Change orders prepared by ENGINEER covering changes in the Work which are required by OWNER, or required because of unforeseen physical conditions or emergencies, or because of uncovering Work found defective, or as provided in paragraphs 11.9 or 11.10, or because of any other claim of CONTRACTOR for a change in the Contract Time or the Contract price which is recommended by ENGINEER.

10.5 If notice of any changes affecting the general scope of the Work or change in the Contract Price is required by the provisions of any bond to be given to the Surety, it will be CONTRACTOR'S responsibility to so notify the Surety, and the amount of each applicable Bond shall be adjusted accordingly. CONTRACTOR shall furnish proof of such adjustment to OWNER.

Exhibit 4-2

The purpose of the clause is to provide a mechanism to accommodate the changes to the work that invariably occur on a job. Not only is it establishing a mechanism, but it is also defining how the contractor will be compensated for cost and time and what steps are required by the contractor if it believes that a change occurred.

In general, if a change arises, the contractor is required to provide written notice of the change within a specified period. The reason that notice is required is to allow all parties to investigate the item that is believed to be a change while the opportunity exists. A time limit is specified so that the owner does not have its rights prejudiced. For example, if a contractor discovered that there was an error in the plans concerning the configuration for the steel reinforcing for a foundation but went ahead and modified the steel to fit, placed the concrete, and then notified the owner, there would be no opportunity for the owner to verify the conflict or the correction made by the contractor. It is possible that the owner could have devised a different, less expensive, solution for the problem. So the lack of timely notice prejudiced the owner's rights. In all cases, it is advisable for the contractor to comply with the provisions stated in the changes clause.

As in the sample clause in Exhibit 4-2, the changes clause may specify several alternative methods for pricing the work. It may be a lump sum, a unit price, or a time and materials basis. The clause further requires that if the owner directs, the contractor is to proceed with the work even if the price has not been resolved. This wording is generally referred to as a continuation of work provision. The reason for this is that you do not want the contract to come to a halt while the two parties attempt to resolve outstanding issues on the change. If the problems cannot be resolved, then the contractor is provided with a mechanism under the "disputes" clause to seek relief.

The precise drafting of the changes clause is very important. If we return to the sample in Exhibit 4-2, we will note that it is silent concerning many aspects of the actual pricing for the change. For instance, if the contractor is submitting a lump-sum cost, is it allowed to include a charge for small tools? What percentage overhead and profit can be applied to the change order pricing? If equipment is used for the work, at what rate is the equipment priced in the change order?

Most clauses do not address these fundamental questions. As a consequence, the parties generally have to resolve them when the first change occurs. It would be far better to resolve these questions up front in the changes clause itself, as in Exhibit 4-3.

Changes

The OWNER may, at any time, without invalidating the Contract, and without notice to the CONTRACTOR'S sureties, make changes within the scope of the Contract which may either increase or decrease the Contract Price and/or Duration. The term "change," as used in the Contract, contemplates a substitution for, an addition to, or deletion of any work or other requirement which is or is not contemplated by the Contract, including, but not limited to the following:

1. A change in the specification (including drawings and design);
2. A change in the method or manner of performance of work, and;
3. A change in the Contract Duration (including acceleration).

A change can only increase or decrease the Contract Price and/or Duration by a written order called a Change Order. Included in the Change Order will be a determination of the increased or decreased Contract Price and/or Duration which will represent full and final satisfaction to the parties for the scope of the change.

The increase or decrease in the Contract Price shall be determined by one or more of the following methods in order of precedence listed below:

1. The applicable Unit Prices;
2. An agreed lump sum amount (equitable adjustment);
3. The actual costs as determined by the Time and Material Basis described herein.

The "applicable unit prices" are those which were agreed upon at the time of the execution of the Contract. Unit Prices may also be negotiated prior to the start of change work which is not defined by the initial contract. Likewise, a Lump sum amount may be negotiated prior to the start of the work of the change.

The Lump Sum amount of equitable adjustment shall include a Breakdown of the CONTRACTOR'S estimated cost. This shall include the cost of labor, material, and equipment required to perform the work. The CONTRACTOR may apply a markup of 15 percent. This markup includes allowance for supervision and clerical. Other allowable costs for a lump sum proposal shall be the same as those specified under the Time and Material Basis.

In the event that the CONTRACTOR and Owner cannot agree upon the value of the equitable adjustment, at the direction of the OWNER, the CONTRACTOR shall commence the change work. All actual costs will be kept by the OWNER and agreed upon on a daily basis. This will be considered the Time and Material Basis.

The Time and Material Basis shall allow reimbursement for the following:

1. *LABOR*
   a. Actual Wages—labor hours, rate as determined by Certified payroll, excluding premium pay, paid to all employees directly engaged in the change below General Foreman level.
   b. Labor Burden—for each craft and class to be established by Certified Payroll as a percent of actual wages paid, including vacation allowance, health and welfare, pension, apprenticeship programs, social security, unemployment insurance, and worker's compensation insurance.
2. *MATERIAL*
   a. Actual material costs as substantiated by invoice at the CONTRACTOR'S net cost including all applicable taxes.
3. *EQUIPMENT*
   a. CONTRACTOR owned equipment—the rental rate as established by the latest edition of the rental rate Blue Book, published by Nielsen Dataquest Inc., of San Jose, California, and shall be calculated as follows:
      Basic Rate—The monthly rate divided by 176 hours per month multiplied by 75 percent.
      Operating Cost—The hourly operating cost, if applicable, shall be used.
      Adjustment—The appropriate area adjustment shall be used before operating costs.
      Overhead & Profit—No markups should be allowed for overhead and profit.
   b. *CONTRACTOR*—Rented Equipment—Equipment rented from an outside rental company shall be paid at the CONTRAC-

TOR's net invoice cost. Transportation will be paid for the equipment if it is applicable solely to the specific change.

4. *OVERHEAD*
   Maximum allowable on labor only 15 percent;
   Maximum allowable on materials, before taxes, 5 percent;
   No markups on equipment allowed.

These allowable overhead markups include supervision and administration of the change above the General Foreman level, such as superintendents, Assistant Superintendents, Engineers, Accountants, Clerks, Timekeepers, Office Manager and all other Field Staff, office supplies, drinking water, temporary heat, light and power; field toilets, costs of services; material and/or equipment not incorporated in the change or directly associated with the Change; small tools, reproduction costs; premiums for all personal, general and public (liability) bonds, and all Home Office costs.

5. *PROFIT*
   Maximum allowable for labor is 5 percent of the sum of actual wages, labor burden and overhead;
   Maximum allowable for materials is 5 percent of the sum of the actual cost and overhead;
   No profit markups for equipment.
6. *SUBCONTRACTOR WORK*
   Shall be quoted in the same manner as for the CONTRACTOR, except the Subcontractor shall not be allowed any Overhead and/or profit on any sub-subcontractors.
   CONTRACTOR shall be allowed a 5 percent markup on the net amount of the total of all acceptable subcontractor costs for the change.

If, in the opinion of the CONTRACTOR, the OWNER has directed the CONTRACTOR to perform change work or the CONTRACTOR encounters a condition that it believes has changed the Contract, the CONTRACTOR must notify the OWNER within 20 days of the occurrence in writing.

This written notice shall take the form of a letter addressed to the OWNER or its representatives which defines the:

1. Applicable contract clauses;
2. Facts surrounding the change work;
3. Reason(s) why the CONTRACTOR believes that a change has occurred;
4. Estimated unit price or lump sum value of the changes, and;
5. Estimated impact to the contract duration.

The OWNER will investigate the conditions and advise the CONTRACTOR if a change has occurred. In the interim the CONTRACTOR and the OWNER will track the costs of the alleged change in accordance with the Time and Material Basis described herein.

In all situations the CONTRACTOR must continue work, unless directed by the OWNER to stop, even prior to the issuance of a Change Order.

Exhibit 4-3

This changes clause is far more specific than the one noted in Exhibit 4-2. All costs are clearly specified to include the rate to be used for equipment and appropriate markups for overhead, profit, and small tools. It also specifies that costs will be tracked on a time and materials basis while a cost is being negotiated.

One final point must be considered concerning the changes clause. Most are worded to the effect that the contractor is to perform change order work after receipt of a signed change order. Yet in the same clause, there is wording to the effect that the owner can direct the contractor to perform the work. To insure that no ambiguity exists, it is advisable that the changes clause describe the mechanism by which extra work can be initiated. This can be a field order, a proceed order, or a similar device. Sample wording is shown below.

In the event that the Contractor and Owner cannot agree upon the value of the change or if any disagreement exists concerning the change, at the written direction of the Owner, the Contractor shall perform the work as directed. . . .

Exhibit 4-4

Differing Site Conditions

Differing site conditions are also considered changed conditions or concealed conditions. Most construction contracts contain a clause that addresses them. Exhibit 4-5 is a sample of this type.

Differing Site Conditions (Changed Conditions)

During the progress of the work, if preexisting subsurface or latent physical conditions are encountered at the site, differing materially from those indicated in the contract, or if preexisting unknown physical conditions of an unusual nature, differing materially from those ordinarily encountered and generally recognized as inherent in the work provided for in the contract, are encountered at the site, the party discovering such conditions shall promptly notify the other party in writing of the specified differing conditions before they are disturbed and before the affected work is performed.

Upon written notification, the Engineer will investigate the conditions and if he/she determines that the conditions materially differ and cause an increase or decrease in the cost or time required for the performance of any work under the contract, an adjustment, excluding loss of anticipated profits, will be made and the contract modified in writing accordingly. The Engineer will notify the Contractor of his/her determination whether or not an adjustment of the contract is warranted.

No contract adjustment which results in a benefit to the contractor will be allowed unless the Contractor has provided the required written notice.

The adjustment will be by agreement with the Contractor. If the parties are unable to agree, the Engineer will determine the amount of the equitable adjustment by establishing the costs or by using force account, and adjust the time as the Engineer deems appropriate. Extensions of time will be evaluated in accordance with Section 1-08.8.

If the Engineer determines differing site conditions (changed conditions) do not exist and no adjustment in costs or time is warranted such determination shall be final as provided in Section 1-05.1.

No claim by the Contractor shall be allowed unless the Contractor has followed the procedures provided in Section 1-04.5 and 1-09.11.

Exhibit 4-5

This clause addresses the situation in which a condition is discovered that was not anticipated based on the bid documents. There are two commonly recognized forms of differing site conditions. A *Type I differing site condition* is a condition that differs materially from that which is represented by the contract documents. For example, the contract may portray the site as being of silt and sand but during excavation, rock is uncovered. A *Type II differing site condition* is a condition of an unusual nature not ordinarily anticipated for that type of work in that area; for example, a sink hole occurring in an area where they are not ordinarily encountered.

The "concealed conditions" clause allows a mechanism for compensation to the contractor should they occur. It should be noted that in the absence of a differing site conditions or concealed conditions clause, then the contractor may be responsible for the costs to perform the work associated with the concealed condition. Generally, it is most reasonable that a clause of this type be included in the contract. If a contractor must start assuming risks for unknown or misrepresented conditions, then the bids for a project will be correspondingly higher.

The owner, however, may be concerned about the affect of a concealed condition on its overall project budget. This quandary can best be resolved by performing more thorough site investigations. An alternative would be to specify unit prices for the potential occurrence of a concealed condition. Some government agencies use a clause termed an "if and where directed" clause. Exhibit 4-6 is an example of this.

5. *"If and Where Directed" Items*

The Proposal form may request bids on one or more Pay Items to be incorporated into the Project "if and where directed" by the Engineer. The Engineer shall have sole discretion in determining whether and to what extent such items will be incorporated into the project. Incorporation of such items into the project shall only be made on written directions of the Engineer. In the absence of written directions, no such items shall be incorporated into the Project and if incorporated shall not be paid for. The engineer may order incorporation of such items at any location within the contract and at any time during the work. These items will not be located on the Plans. The estimated quantities set out in the Proposal for such items are presented solely for the purpose of obtaining a representative bid price. The actual quantities employed may be only a fraction, or many times the estimated quantity. The Contractor shall make no claim for additional compensation because of any increase, decrease, or elimination of such items.

Exhibit 4-6

In our previous example of the rock appearing where none was expected, the if and where directed items could have included a quantity for rock excavation with a bid unit price. In that fashion, the owner is setting the price up front and reducing the risk should the problem occur. A point of caution must be noted. The use of an if and where directed clause is not intended as a substitute for good site studies, accurate quantity calculations, or well-written contract documents. It is solely intended as a safety valve to prevent problems and reduce risk to all the parties.

Scheduling Requirements

The contract should have a well-written requirement for a schedule for the project. Chapter 5 addresses scheduling in detail. Suffice it to say, if the contract does not require a schedule, one may not be submitted for the project.

Unfortunately, most construction contracts do not have well-structured requirements for project schedules. It is common to see clauses of the type shown below.

GP-8.04 Progress Schedule

A. Within 30 days after notice to proceed, the Contractor shall furnish the procurement officer a "Progress Schedule" showing the proposed order of work and indicating the time required for the completion of the work. Said progress schedule shall be used to establish major construction operations and to check on the progress of the work. The contractor shall submit revised progress schedules as directed by the procurement officer.

B. If, the Contractor fails to submit the progress schedule within the time prescribed, or the revised schedule within the requested time, the procurement officer may withhold approval of progress payment estimates until such time as the Contractor submits the required progress schedules.

C. If in the opinion of the procurement officer, the Contractor falls significantly behind the approved progress schedule, the Contractor shall take any and all steps necessary to improve his progress. To accomplish this action may require the Contractor to increase the number of shifts, initiate or increase overtime operations, increase days of work in the work week, or increase the amount of construction plants, or all of them. The procurement officer may also require the Contractor to submit for approval supplemental progress schedules detailing the specific operational changes to be instituted to regain the approved schedule, all without additional cost to the Administration.

D. Failure of the Contractor to comply with the requirements of the procurement officer under this provision shall be grounds for determination by the procurement officer that the Contractor is not prosecuting the work with such diligence as will insure completion within the time specified. Upon such determination, the procurement officer may terminate the Contractor's right to proceed with the work, or any separable part thereof, in accordance with GP-8.08 of these General Provisions.

Exhibit 4-7

Just a brief review of this clause shows that the contractor could submit a minimal bar chart and satisfy the requirements of the contract. During the planning phase, the owner must decide what kind of project schedule is needed, how it will be effected, and then draft the corresponding clause. Exhibit 4-8 is an example of a clause that more clearly spells out what the owner desires.

Sample Scheduling Specification
(Owner Performs Mechanics)

The construction of this project will be planned and recorded with a conventional Critical Path Method (CPM) schedule. The schedule shall be used for coordination, monitoring, and payment of all work under the contract including all activity of subcontractors, vendors, and suppliers.

CONTRACTOR is responsible for preparing the initial schedule in the form of an activity on arrow diagram. Owner will provide a scheduling expert to work with CONTRACTOR in preparing the initial schedule. OWNER is responsible for providing computer processing of the scheduling data provided by CONTRACTOR. All costs incurred by CONTRACTOR in preparing the schedule shall be borne by CONTRACTOR as a part of its responsibility under this contract.

A. *60-Day Preliminary Schedule*

Before proceeding with any work on site, CONTRACTOR shall prepare, submit, and receive OWNER'S approval of a 60-Day Preliminary Schedule. This schedule shall provide a detailed breakdown of activities scheduled for the first 60 days of the project and shall include mobilization, submittals, procurement, and construction.

No contract work may be pursued at the site without an approved 60-Day Preliminary Schedule or an approved CPM schedule.

B. *Preparation Of Initial Schedule*

Within 10 calendar days of the contract award, CONTRACTOR shall meet with the OWNER'S expert to begin developing the initial schedule. Within 30 days of the Notice to Proceed, CONTRACTOR will complete development of the initial schedule and present to OWNER an activity on arrow diagram depicting its schedule for computer processing by OWNER.

Following computer processing and within 14 calendar days of submission of the diagram, OWNER and CONTRACTOR shall meet for joint review, correction, and adjustment of the schedule. The construction time, as determined by the schedule, for the entire project or any milestone shall not exceed the specified contract time. In the event that any milestone date or contract completion date is exceeded in the schedule, logic and/or time estimates will be revised.

After any changes in the logic and/or time estimates have been agreed upon, another computerized schedule will be generated. The process will be repeated, if necessary, until the schedule meets all contractual requirements. However, the schedule must be finalized within 60 days of the Notice to Proceed. Failure to finalize the schedule by that date will result in withholding all contract payments until the schedule is finalized.

Once the initial schedule has been finalized and is within contract requirements, CONTRACTOR shall submit a signed copy of the schedule to OWNER.

C. *Schedule Requirements*

All activity on arrow diagrams shall include:

1. Activity nodes
2. Activity Description
3. Activity Duration

The activity on arrow diagram shall show the sequence and interdependence of all activities required for complete performance of all items of work under this contract, including shop drawing submittals and approvals and fabrication and delivery activities. All network "dummies" are to be shown on the diagram.

No activity duration shall be longer than 15 work days without OWNER'S approval.

OWNER reserves the right to limit the number of activities on the schedule.

The activities are to be described so that the work is readily identifiable and the progress of each activity can be readily measured. For each activity CONTRACTOR shall identify the trade or subcontractor performing the work, the duration of the activity in work days, the manpower involved by trade, the equipment involved, the location of the work, and a dollar value of the activity. The dollar value assigned to each activity is to be reasonable and based on the amount of labor, materials, and equipment involved. When added together the dollar value of all activities are to equal the contract price.

CONTRACTOR shall also provide the following information: work days per week, holidays, number of shifts per day, number of hours per shift, and major equipment to be used.

Any activity on arrow diagram submitted by the CONTRACTOR or prepared by the OWNER'S expert, may either be hand drawn or computer plotted. Regardless of the type of diagram, the network must be legible, readable, and understandable. Network diagram will be on standard D size sheets (24″ × 36″) and not a continuous diagram.

Any network diagram submitted by the Contractor must include one reproducible sepia and three copies.

For both the initial schedule and all updates, the OWNER'S expert will provide the following:

1. Computerized sorts by:
   I-J
   Total Float
   Early Start

Area Sort
Trade Responsibility

2. 60-day look ahead bar charts by early start.
3. A narrative explaining progress to date on the project, work required in the succeeding update period, a description of the critical path, and comments concerning potential problem areas.
4. CONTRACTOR will be provided with four copies of each of the above.

D. *Schedule Updates And Progress Payments*

Job site progress meetings will be held monthly by OWNER and CONTRACTOR for the purpose of updating the project work schedule and determining the appropriate amount of partial payment due CONTRACTOR. Progress will be reviewed to verify finish dates of completed activities, remaining duration of uncompleted activities, and any proposed logic and/or time estimate revisions. It is CONTRACTOR'S responsibility to provide OWNER with the status of activities at this progress meeting. The CONTRACTOR will report progress on a daily basis in accordance with the attached form. OWNER will process schedule updates based on this information once it has been verified.

CONTRACTOR will submit revised activity on arrow diagrams for the following: delay in completion of any critical activity; actual prosecution of the work which is, as determined by OWNER, significantly different than that represented on the schedule; or the addition, deletion, or revision of activities required by contract modification. The contract completion time will be adjusted only for causes specified in this contract.

As determined by CPM analysis, only delays in activities which affect milestone dates or contract completion dates will be considered for a time extension.

If CONTRACTOR does seek a time extension of any milestone or contract completion date, it shall furnish documentation as required by OWNER to enable OWNER to determine whether a time extension is appropriate under the terms of the contract.

It is understood by OWNER and CONTRACTOR that float is a shared commodity.

The principals involved and terms used in this section are as set forth in the Associated General Contractors of America publication, "The Use of CPM in Construction, A Manual for General Contractors and the Construction Industry," Copyright 1976.

Exhibit 4-8

This clause requires a CPM schedule, specifies timeframes for its development and submission, details the exact submissions required, and establishes a mechanism to enforce the clause. If the schedule is not submitted, all payments are withheld. In this particular clause, the owner wants to administer the schedule with the assistance of its own scheduling consultant as opposed to relying solely on the contractor. Regardless of the approach, the clause must have a level of detail similar to that shown in Exhibit 4-8.

Time Extensions

While the changes clause may refer to an extension of time for additional work, the mechanics of a time extension are normally addressed in a separate clause. The drafting of this type of clause depends on how time extensions are to be evaluated and granted. As amazing as it may seem, many clauses gauge time extensions on the dollar value of extra work performed. A clause of this type is shown in Exhibit 4-9.

(Example 1) Time Extensions: Should the quantities specified in the contract be increased or should extra work be authorized by the Engineer, an extension of the contract time shall be allowed if requested by the Contractor. The extension of contract time will be based on the percentage increase of the dollar value of the increased quantities or extra work. Based on the dollar percentages, that percentage applied to the original duration of the contract will be the maximum time allowed.

(Example 2) Time Extensions: The number of days for performance allowed in the Contract as awarded is based on the proposal quantities. If satisfactory fulfillment of the Contract with extensions and increases authorized under Sections on Variations in Estimated Quantities and Changes as outlined in the General Provisions and Supplemental Agreements shall require the performance of work in greater quantities that those set forth in the Proposal, the Contract time allowed for performance shall be adjusted in an equitable manner based on the quantities, costs and the nature of the work involved.

Exhibit 4-9

From a scheduling and a critical path perspective, this manner of calculating time makes no sense. For example, an owner could approve a change order for $100,000 for the cost associated with the upgrade of some equipment to be installed in a facility. If the contract amount was $2,000,000 and the duration was originally 200 days, a change order for $100,000 would mean that a time extension would be due for ten days. In reality, project time might not be affected at all.

A more realistic approach to the "time extensions" clause is to tie additional time to the schedule and the updates of that schedule. Exhibit 4-10 is an example of this type of clause.

Determination of Extension of Contract Time

The number of days allowed for completion of the work included in the Contract will be stated in the Proposal and contract, and will be known as the "Contract Time."

If the Contractor experiences delays to the completion of the project which are unforeseeable and without the fault or negligence of the Contractor (excusable delays), the time allowed for performance of the work may be extended in accordance with the following:

(a) For delays caused by acts of god, acts of the public enemy, fires, floods, epidemics, quarantine restrictions, strikes, freight embargoes, or unusually severe weather, the Contractor may be granted an extension of time but no additional compensation provided that the Contractor, within five days from the beginning of any such delay, notifies the Engineer in writing of the nature of the delay.

(b) For delays caused by the Department the Contractor may be granted an extension of time and additional compensation in accordance with Article XXXX provided that the Contractor, within five days from the beginning of any such delay, notifies the Engineer in writing of the nature of the delay.

Any requests for time extensions must be supported by a progress schedule analysis in accordance with Article 108.03 (Progress Schedules).

As determined by CPM analysis, only delays in activities which affect milestone dates or contract completion dates will be considered for a time extension.

When final acceptance is made by the Engineer as prescribed in Subsection 105.16, the daily time charge will cease.

Exhibit 4-10

As described, a time extension is granted only if the schedule demonstrates that the critical path of the project was delayed and that the delay could not be mitigated by reasonable rescheduling or resequencing.

There is a second portion of the time extension clause that is of particular significance. That portion deals with whether a delay or time extension also entitles the contractor to additional compensation. Normally, delays that have entitlement to additional compensation are referred to as compensable delays. Delays where the contractor is entitled only to time are termed noncompensable time extensions. The contract must specify what constitutes a noncompensable time extension. This subject will be addressed in more detail in Chapter 6 under "Delays." For the present, Exhibit 4-11 is an example of a clause which differentiates among the various forms of delays that might occur and indicates which are compensable.

Time Extensions/Delays

If the Contractor experiences delays to the completion of the project which are unforeseeable and without the fault or negligence of the Contractor (excusable delays), the time allowed for performance of the work may be extended in accordance with the following:

(a) For delays caused by acts of god, acts of the public enemy, fires, floods, epidemics, quarantine restrictions, strikes, freight embargoes, or unusually severe weather, the Contractor may be granted an extension of time but no additional compensation provided that the Contractor, within five days from the beginning of any such delay, notifies the Engineer in writing of the nature of the delay.

(b) For delays caused by the Department, the Contractor may be granted an extension of time and additional compensation in accordance with Article XXXX provided that the Contractor, within five days from the beginning of any such delay, notifies the Engineer in writing of the nature of the delay.

Exhibit 4-11

Disputes

A disputes clause is a "seat belt" clause. Hopefully, we all wear seat belts when riding in an automobile. At the same time we trust that we will never really need them. The analogy applies to the disputes clause. We need to have a vehicle in place for resolving disputes, but if all goes well, we won't ever have to use it.

The disputes clause should clearly define the method that will be employed to resolve disputes should they arise during the course of the project. Obviously, there are many options that could be used, including litigation, arbitration, dispute review boards, administrative boards, mini trials, neutrals, and mediation. Commonly you will see a choice of either arbitration or litigation. Recently, though, many contracts are adopting procedures for alternative dispute resolution. The reason for this is to stay out of the courts and try to avoid the associated legal fees.

When choosing a method for resolving disputes, evaluate the relative advantages and disadvantages of each. Based on that evaluation, choose the best method or combination of methods for your particular situation. Exhibit 4-12 is an example of a disputes clause in which the method chosen is litigation, except for disputes of less than $100,000, in which case they are handled by arbitration.

Resolution of Claims and Disputes

Any controversy or Claim arising out of or related to the Contract, or the breach thereof which has an aggregate amount of less than $100,000, shall be settled by arbitration in accordance with the Construction Industry Arbitration Rules of the American Arbitration Association, and judgement upon the award rendered by the arbitrator or arbitrators may be entered in any court having jurisdiction thereof, except controversies or Claims relating to aesthetic effect and except those waived as provided for in Subparagraph 4.3.5. Such controversies or Claims upon which the Architect has given notice and rendered a decision as provided in Subparagraph 4.4.4 shall be subject to arbitration upon written demand of either party. Arbitration may be commenced when 45 days have passed after a Claim had been referred to the Architect as provided in Paragraph 4.3 and no decision has been rendered. Any Claims in excess of $100,000 shall be decided in accordance with the laws of (State) in the Circuit Court of (define).

Exhibit 4-12

Exhibit 4-13 is an example of a disputes clause establishing a disputes board that must be utilized before any litigation can occur. This is generally a good idea. To date, disputes boards have demonstrated an outstanding track record in keeping parties out of court.

DISPUTES RESOLUTION

In order to assist in the resolution of disputes or claims arising out of the work of this project, the State has provided for the establishment of a Disputes Review Board, hereinafter called the "BOARD." The BOARD has been added to the disputes resolution process to be brought into play to a formal adoption of position or filing of litigation by either party.

A. Disputes

Disputes, as used in this Section, will include disagreements, claims, counterclaims, matters in question, and differences of opinion between the State and Contractor.

1. On matters related to the work and to change order work including:
   a. Interpretation of Contract Documents
   b. Costs
   c. Time for performance.
2. And on other subjects mutually agreed by the State and Contractor to be of concern of the BOARD.

B. Resolution Procedure

The following procedure shall be used for dispute resolution:

1. If the Contractor objects to any decision or order of the Engineer, the Contractor shall request, in writing, written instructions from the Engineer.
2. The Engineer shall respond, in writing, to the Contractor's written request within 15 calendar days.
3. Within 30 calendar days after receipt of the Engineer's written instructions, the Contractor shall, if the Contractor still objects to such instructions, file a written protest with the Engineer, stating clearly and in detail the basis of the objection that:
   a. Cites contract provisions that support the protest,
   b. Estimates the dollar cost, if any, of the protested work, and
   c. Estimates the amount of additional time incurred, if any.
4. The Engineer will consider any written protest and make a decision on the basis of the pertinent contract provisions and facts and circumstances involved in the dispute. The decision will be furnished in writing to the Contractor within 60 calendar days. This decision shall be final and conclusive on the subject unless a written appeal is filed by the Contractor. Should the Contractor appeal the Engineer's decision, the matter can be referred to the BOARD by either the State or the Contractor.
5. The Contractor's appeal for review must be instituted within 30 calendar days of the date of receipt of the Engineer's decision.
6. The Contractor and the State shall each be afforded an opportunity to be heard by the BOARD and to offer evidence. Either party furnishing any written evidence or documentation to the BOARD must furnish copies of such information to the other party a minimum of 15 calendar days prior to the date the BOARD sets to convene the hearing for the dispute. Either party shall produce such additional evidence as the BOARD may deem

necessary to an understanding and determination of the dispute and furnish copies to the other party.

7. The BOARD's recommendations toward resolution of a dispute will be given in writing to both the State and the Contractor. The recommendations will be based on the contract provisions and the actual costs and time incurred.
8. Within 30 calendar days of receiving the BOARD's recommendations, both the State and the Contractor shall respond to the other in writing signifying that the dispute is either resolved or remains unresolved.
9. Although both parties should place weight upon the BOARD's recommendations, the recommendations are not binding. Either party may appeal a recommendation of the BOARD to them for reconsideration. However, reconsiderations shall only be allowed when there is new evidence to present.
10. If the State and the Contractor are able to resolve their dispute with the aid of the BOARD's recommendations, the State will promptly process any contract changes.
11. In the event the BOARD's recommendations do not resolve the dispute, all records, and written recommendations including any minority reports, will not be admissible as evidence in any subsequent litigation.

C. Litigation

1. Submittal of dispute to the BOARD shall be a condition precedent to filing for litigation in a court of law unless the State and the Contractor have agreed to defaulting to Section 1-09.11(2), Claims.
2. Claims, counterclaims, disputes, and other matters in question between the State and Contractor that are not resolved will be decided in the Superior Court of Thurston County, Washington, which shall have exclusive jurisdiction and venue over all matters in questions between the State and the Contractor.
3. The Contract Documents shall be interpreted and construed in accordance with the laws of the State.

D. Purpose and Function of the BOARD

The BOARD will be an advisory body created to assist in the resolution of claims, disputes, or controversy between the Contractor and the State in order to prevent construction delay and possible court litigation.

The BOARD will consider disputes referred to it, and furnish recommendations to the State and Contractor to assist in the resolution of the differences between them. The BOARD will essentially be making nonbinding findings and recommendations and provide special expertise to assist and facilitate the resolution of disputes.

E. BOARD Members

The BOARD shall consist of one member selected by the State and one member selected by the Contractor, with these two members to select the third member. The first two members shall be mutually acceptable to both the State and the Contractor. If one or both of the two members selected are not acceptable to the State or Contractor, another selection shall be made.

The State and Contractor shall each select their member and negotiate an agreement with their respective BOARD members within the first 60 calendar days after becoming aware of the contract. These negotiated agreements shall include a clause that requires the respective selected members to immediately pursue selection of the third member in accordance with Section IIA of the Disputes Review Board Three Party Agreement in the Appendix of these Special provisions.

In the event of an impasse in selection of the third member, either the State or the Contractor or both may appeal to the Thurston County Superior Court for selection of a third member by the court from a list or lists submitted to the court by the State and or the Contractor. An impasse shall be considered to have been reached if the two members appointed by the State and Contractor for the BOARD have been unable to appoint the third member in a period of 60 calendar days after the approval of the last two such members.

In case a member of the BOARD needs to be replaced, the replacement member will be appointed in the same manner as the replaced member was appointed. The appointment of a replacement BOARD member will begin promptly upon determination of the need for replacement and shall be completed within 30 calendar days. The Three Party Agreement will be amended to reflect the change of a BOARD member.

Service of a BOARD member may be terminated at any time with not less than 30 calendar days notice as follows:

1. The state may terminate service of the State appointed member.
2. The Contractor may terminate service of the Contractor appointed member.
3. The third member's services may be terminated only by agreement of the other two members.
4. By resignation of the member.
5. Termination of a member will be followed by appointment of a substitute as specified above.

No member shall have a financial interest in the contract, except for payments for services on the BOARD. No member shall have been employed by either party within a period of two years prior to award of this contract; except that, service as a member of other Disputes Review

Boards on other contracts will not preclude a member from serving on the BOARD for this contract.

The BOARD members will be especially knowledgeable in the field of construction of the type covered by the Contract documents and shall discharge their responsibilities impartially and independently considering the facts and conditions related to the matters under consideration and the provisions of the Contract Documents.

F. BOARD Operation

The BOARD will formulate its own rules of operation. In order to keep abreast of the construction development, the members shall regularly visit the project, keep a current file, and regularly meet with the other members of the BOARD and with representatives of the State and the Contractor.

The frequency of these visits shall be agreed between the State, the Contractor, and the BOARD.

For further description of work, responsibilities and duties of the BOARD, and the State's and Contractor's obligations and responsibilities with respect to each other and to the BOARD, see the Disputes Review Board Three Party Agreement in the Appendix of these Special Provisions.

Exhibit 4-13

One note of caution should be sounded. If a construction contract specifies a particular method for disputes resolution such as arbitration, some consideration should be made to have similar wording in the design contract. Consider the example of the project where the owner and contractor became enmeshed in a dispute and the method of resolution was arbitration. The nature of the dispute involved the designer and its actions on the project. The design contract required litigation as a means of settling disputes. As a consequence, the owner and contractor were unsuccessful in having the dispute heard in a common forum. Instead, there were two separate actions and additional expense.

Liquidated Damages

Clauses that address delays caused by the contractor or subcontractor are common in construction contracts. The most common form is a liquidated damage clause. A sample of this type of clause is shown in Exhibit 4-14.

2. Liquidated Damages

In case of failure on the part of the Contractor to complete the work within the times fixed in the contract or any extensions thereof, the Contractor shall pay to the Government as liquidated damages, pursuant to the clause of this contract entitled "Termination for Default—Damages for Delay—Time Extensions," the sum of $400.00 for each day of delay in completing all work.

Exhibit 4-14

If the project is delayed due to the fault of the contractor, the owner has the right to be compensated for the additional costs that it incurs. In drafting the contract, the owner has the option of seeking actual damages if a delay occurs. If so, there is no need for a liquidated damage clause. In many cases, particularly in public works, the damages that the owner experiences because of a delay are difficult to quantify. For this reason, a liquidated damage clause is used. In other words, liquidated damage clauses are used where it is difficult or impossible to quantify the exact amount of damages.

Liquidated damage clauses are enforced by the courts. Caution should be exercised, however, when drafting such clauses. The amount of damages specified should be based on a reasonable estimate of the potential extra costs if a delay occurs. The amount should not just be selected randomly. The purpose of the clause is to reimburse the owner, not penalize the contractor. You should consult with a construction attorney concerning the ramifications of these clauses, but, in general, you should know that a liquidated damage clause that appears to be a penalty, will not be enforced.

Closely related to liquidated damage clauses are incentives/disincentives clauses. In the past these have been referred to as "bonus/penalty" clauses. The nomenclature changed because of court decisions concerning the unenforcability of penalties. An example of this type of clause is shown in Exhibit 4-15.

The following incentive/disincentive will apply to the work listed above. The incentive payment will be limited to a maximum of $1,050,000.

a. *Disincentive—Failure to Complete on Time*—If the contractor fails to complete the work item on or before the calendar days specified or on or before the authorized extension thereof without disincentives, the contractor will be charged, for each calendar day that the work shall remain uncompleted, the sum set forth below.

b. *Incentive—Early Completion*—If the contractor completes the work items before the calendar days specified or before the authorized extension thereof not subject to disincentives, the contractor will be reimbursed, for each calendar day from the time the work is completed to the contract time specified or authorized extension thereof, the sum set forth below.

c. Incentives and disincentives will be paid for as follows:
   1. *Incentive*—for each day of underrun, payment will be increased by $15,000 per calendar day.
   2. *Disincentive*—For each day of overrun, payment will be decreased by $15,000 per calendar day.

The entire project shall be completed by June 1, 1990.

The contractor is alerted to the necessity for timely fabrication and delivery of critical materials needed to complete the above items of work. Extensions of time will not be granted for delays in delivery of critical materials, except for delays which can be shown to be industrywide.

No extensions of time will be granted for labor disputes unless it can be shown that such disputes are industrywide.

No extensions or incentive/disincentive dates will be granted for weather conditions.

Exhibit 4-15

This type of clause should be approached in the same manner as the liquidated damage clause. The amount of the disincentive should be based on a reasonable estimate at the time that the clause is drafted. The incentive should be based on an estimate of the benefits that the owner will realize if the project is completed early. The amount of the incentive does not have to be equal to the amount for the disincentive.

There is a common misconception in the industry that a liquidated damage clause cannot be enforced without a corresponding bonus or incentive. There is no requirement that an incentive be included.

Finally, it should be noted that liquidated damages or incentive/disincentive clauses can be established not only for final completion of the project, but also for milestone completion dates for various portions of the work. This is often an astute practice for the general contractor that uses subcontractors.

Termination

The "termination" clause in a construction contract is a necessity. In essence, it states that should the contractor be found in default of the contract, the owner has the right to terminate the contract for cause. An example of this type of clause is shown below.

Default

(a) If the Contractor refuses or fails to prosecute the work or any separable part, with the diligence that will insure its completion within the time specified in this contract, including any extension, or fails to complete the work within this time, the Government may, by written notice to the Contractor, terminate the right to proceed with the work (or separable part of the work) that has been delayed. In this event, the Government may take over the work and complete it by materials, appliances, and plant on the work site necessary for completing the work. The Contractor and its sureties shall be liable for any damages to the Government resulting from the Contractor's refusal or failure to complete the work within the specified time, whether or not the Contractor's right to proceed with the work is terminated. The liability includes any increased costs incurred by the Government in completing the work.

(b) Contractor's right to proceed shall not be terminated nor the Contractor charged with damages under this clause, if—

(1) The delay in completing the work arises from unforeseen causes beyond the control and without the fault or negligence of the Contractor. Examples of such clauses include (i) acts of God or of the public enemy, (ii) acts of the Government in either its sovereign or contractual capacity, (iii) acts of another Contractor in the performance of a contract with the Government, (iv) fires, (v) floods, (vi) epidemics, (vii) quarantine restrictions, (viii) strikes, (ix) freight embargoes, (x) unusually severe weather, or (xi) delays of subcontractors or suppliers at any tier arising from unforeseeable causes beyond the control and without the fault or negligence of both the Contractor and the subcontractor suppliers; and

(2) The Contractor, within 10 days from the beginning of any delay (unless extended by the Contracting Officer), notifies the Contracting Officer in writing of the causes of delay. The Contracting Officer shall ascertain the facts and the extent of the delay. If, in the judgement of the Contracting Officer, the finding of fact warrant such actions, the time for completing the work shall be extended. The findings of the Contracting Officer shall be final and conclusive on the parties, but subject to appeal under the Disputes clause.

(c) If, after termination of the Contractor's right to proceed, it is determined that the Contractor was not in default, or that the delay was excusable, the rights and obligations of the parties will be the same as if the termination had been issued for the convenience of the Government.

(d) The right and remedies of the Government in this clause are in addition to any other rights and remedies provided by law or under this contract.

Exhibit 4-16

The termination clause should specify what constitutes default of contract. The more general the clause, the more likely it is that problems will arise if its enforcement is attempted. The exact prerequisites for default are a decision on the part of the drafter of the contract. The most common causes cited include bankruptcy, failure to maintain progress, and failure to follow the directives of the owner. The specific causes inserted depend on the nature of the project. For instance, one general contractor concerned about a tight project schedule included the requirement that a subcontractor was in default if it fell more than ten days behind the current CPM schedule for the project.

The termination clause should contain a provision that the termination reverts to a termination for convenience if it is later determined that the contractor was not in default. This precludes a wrongfully terminated contractor from seeking damages for lost profits, opportunity costs, and so on.

The "termination for convenience" clause is the second form of termination clause that the construction contract should include. An example of this type of clause is shown in Exhibit 4-17.

Termination by the Owner for Convenience

(a) The Owner may terminate performance of work under this contract in whole or, from time to time, in part, if the Owner determines that a termination is in the Owner's interest. The Owner shall terminate by delivering to the Contractor a Notice of Termination specifying the extent of termination and the effective date.

(b) After receipt of a Notice of Termination, and except as directed by the Owner, the Contractor shall immediately proceed with the following obligations, regardless of the status of determining or adjusting any amounts that may be due under this clause:

(1) Stop work as specified in the notice.

(2) Place no further subcontracts or orders (referred to as subcontracts in this clause) for materials, services, or facilities, except as necessary to complete the continued portion of the contract.

(3) Terminate all subcontracts to the extent they relate to the work terminated.

(4) Assign to the Owner, as directed by the Owner, all rights, title, and interest of the Contractor under the subcontracts terminated, in which case the Owner shall have the right to settle or to pay any termination settlement proposal arising out of those terminations.

(5) With approval or ratification to the extent required by the Owner, settle all outstanding liabilities and termination settlement proposals arising from the termination of subcontracts; the approval or ratification will be final for purposes of this clause.

(6) As directed by the Owner, transfer title and deliver to the Owner (i) the fabricated or unfabricated parts, work in process, completed work, supplies, and other material produced or acquired for the work terminated, and (ii) the completed or partially completed plans, drawings, information, and other property that, if the contract had been completed, would be required to be furnished to the Owner.

(7) Complete performance of the work not terminated.

(c) After termination, the Contractor shall submit a final termination settlement proposal to the Owner in the form and with the certification prescribed by the Owner. The Contractor shall submit the proposal promptly, but no later than 1 year from the effective date of termination, unless extended in writing by the Owner upon written request of the Contractor within this 1-year period.

(d) Subject to paragraph (c) above, the Contractor and the Owner may agree upon the whole or any part of the amount to be paid because of the termination. The amount may include a reasonable allowance for profit on work done. However, the agreed amount, whether under this paragraph or paragraph (e) below, exclusive of costs shown in subparagraph (e)(3) below, may not exceed the total contract price as reduced by (a) the amount of payments previously made and (2) the contract price of work not terminated. The contract shall be amended, and the Contractor paid the agreed amount. Paragraph (e) below shall not limit, restrict, or affect the amount that may be agreed upon to be paid under this paragraph.

(e) If the Contractor and Owner fail to agree on the whole amount to be paid the Contractor because of the termination of work, the Owner shall pay the Contractor the amounts determined as follows, but without duplication of any amounts agreed upon under paragraph (d) above;

(1) For contract work performed before the effective date of termination, the total (without duplication of any items) of—

(i) The cost of this work

(ii) The cost of settling and paying termination settlement proposals under terminated subcontracts that are properly chargeable to the termination portion of the contract; and

(iii) A sum, as profit on (i) above, determined by the Owner to be fair and reasonable; however, if it appears that the Contractor would have sustained a loss on the entire contract had it been completed, the Owner shall allow no profit under this subdivision (iii) and shall reduce the settlement to reflect the indicated rate of loss.

(2) The reasonable costs of settlement of the work terminated, including—

(i) Accounting, legal, clerical, and other expenses reasonably necessary for the preparation of termination settlement proposals and support data;

(ii) The termination and settlement of subcontracts; and

(iii) Storage, transportation, and other costs incurred, reasonably necessary for the preservation, protection, or disposition of the termination inventory.

(f) In arriving at the amount due to Contractor, there shall be deducted—

(1) All unliquidated advance or other payments to the Contractor under the terminated portion of this contract;

(2) Any claim which the Owner has against the Contractor under this contract; and

(3) The agreed price for, or the proceed of sale of materials, supplies, or other things acquired by the Contractor or sold under the provisions of this clause and not recovered by or credited to the Owner.

Exhibit 4-17

This clause allows the owner to terminate the contractor's work for reasons other than default. For instance, during the construction of the project it might be determined that the project is no longer economically feasible. Therefore, the most cost-effective decision is to stop the job. Similarly, an owner may have a contractor whose work is not satisfactory but is not bad enough to be in default. The termination for convenience clause enables the owner to end the contract and complete the work by an alternate means. With this clause, the owner specifies the method of payment and precludes potential disputes concerning the amount due the terminated contractor. Since the clause normally specifies that the contractor is due its actual costs for the work performed, it is advisable to include the provision that the owner has the right to audit to verify the costs that are to be paid.

Exculpatory Language

Because of the increasingly litigious nature of construction projects, many contracts have adopted a highly protective and one-sided approach through the use "exculpatory" clauses, which protect one party by putting the risk on the other party to the contract. In essence, the clause exculpates or excuses one party from a certain risk.

The most common form of exculpatory clause used is one known as a no damage for delay clause. An example of this is shown below.

ART. 95—Extension of Time for Completing the Work

If the Contractor is delayed in completion of the work under the Contract by any act or neglect of the Owner or of any other Contractor employed by the Owner, or by changes in the work, or by any priority or allocation order duly issued by the Federal Government, or by any unforeseeable cause beyond the control and without the fault or negligence of the Contractor, excluding, but not restricted to, acts of God or of the public enemy, fires, floods, epidemic, quarantine restrictions, strikes, freight embargoes, and abnormally severe weather, or by delays of subcontractors or suppliers occasioned by any of the causes described above, or by delay authorized by Engineer for any cause which the Engineer shall deem justifiable, then:

For each day of the delay in the completion of the work so caused, the Contractor shall be allowed one day additional to the time limitation specified in the Contract, it being understood and agreed that the allowance of same shall be solely at the discretion and approval of the Owner.

No such extension of time shall be made for any delay unless the Contractor, within 3 days after the beginning of the delay, shall have informed the Owner in writing of the nature of the delay, its cause, and its limited duration. The Owner will ascertain the facts regarding the delay and notify the Contractor within a reasonable time of its decision in the matter.

The Contractor shall use all honorable and reasonable means to prevent strikes, to avoid violations of labor agreements or other actions calculated to create dissatisfaction with working conditions. Should strikes occur, it shall make all proper and reasonable efforts that effect early settlement and resumption of the work. Should collusion by the Contractor be proven in the case of strikes or lockouts, then no extension of time for completion of the Contract will be given. Burden of proof in this case shall rest entirely with the Contractor.

No claim for damages or any claim other than for extensions of time as herein provided shall be made or asserted against the Owner by reasons of any delays caused by the reasons hereinabove mentioned.

Exhibit 4-18

This clause places the burden of any form of delay on the contractor. It notes that the owner will grant a time extension but pay no additional costs caused by the delay. Clauses of this type are enforceable but, as always, there are exceptions. The reader should consult with qualified construction counsel or refer to the appropriate texts in the bibliography.

Using a clause of this type obviously shifts the risk to the contractor. Consequently, either the bids will be increased because of it, or if a delay does occur that is caused by the owner, protracted and costly litigation may result. It is the author's opinion that exculpatory language is not the preferred method to reduce risk. It either increases the chances of disputes or it drives the bid prices up. This is true of all forms of exculpatory clauses. Given the risks and high costs associated with delays, it is more prudent to specify exactly what the owner will pay for a delay as opposed to using a no damage for delay clause.

The next most common form of exculpatory clause involves site conditions. These clauses can take many forms. An example is shown in Exhibit 4-19.

INFORMATION NOT GUARANTEED

All information given in the Contract documents, including drawings, relative to borings and materials encountered, ground water, subsurface conditions, and existing pipes and other structures is from the best source at present available to the Owner. All such information is furnished only for the convenience of bidders.

It is understood and agreed that the Owner does not warrant or guarantee that the material, conditions, and pipes and other structures encountered during construction will be the same as those indicated by the boring samples or by the information given on the drawings. The bidder must satisfy himself regarding the character, quantities, and conditions of the various materials and the work to be done.

SUBSURFACE STRUCTURES

The approximate location of certain existing subsurface pipes, utilities, and structures is indicated on the drawings. The extent, character, and location of said items are not guaranteed.

Exhibit 4-19

As expressed in the clauses, the owner disclaims the accuracy of all of the site conditions as presented in the contract. Hence, the contractor is forced to assume the risk of any form of differing site condition. Once again, either the bids will be higher to allow for contingencies for this or a dispute may occur. This has not reduced the risk. Instead, it has drawn battle lines for the parties involved.

The use of a site disclaimer defies logic. For example, the reason that an owner performs site investigation is to determine information that is vital to the basic project design. The owner relies on that information in formulating the design, but, in the same breath, states that the contractor cannot rely on it. This seems both contradictory and unfair.

To reduce risk of differing site conditions, it makes more sense to make more detailed site studies and use "if and where "directed" clauses. If there are significant concerns regarding the site conditions, the owner might utilize separate contracts for the site/foundation work and the remainder of the construction. In this manner, the owner can limit its risk and potential extra costs.

Exculpatory clauses can go to amazing extremes. Exhibit 4-20 is an example of this.

ART. 26—Acceptance of Contract Documents

The Contractor admits and agrees that he is satisfied with the Contract Documents and agrees that he will at no time dispute or complain that there was any misunderstanding or any error in regard to the materials and equipment to be furnished, their amounts and quantities, and the work to be done under this Contractor or in regard to the amount of compensation to be paid therefor; and he further covenants and agrees to completely execute and perform his Contract and to fully complete the said work or improvements to the satisfaction of the Owner and to strictly comply with these Contract Documents and not to ask or demand, sue for or recover any further or extra compensation beyond the Contract price. It is intended that the Contract price with adjustments as noted in the Proposal shall be the sole and only compensation to the Contractor for the full and complete performance of this Contract, and the full completion of said Contract. It is also understood and agreed that the price to be paid includes payment for all labor, material, tools, equipment, tests, guarantees, and permit therefor.

The Contractor accepts the Contract Documents as complete and accurate and agrees that there is no conflict therein with permissible trade practices or methods. Any objections to the Contract documents that the Contractor may have must be called to the Engineer's attention and the matter resolved before submitting his Proposal.

The Contractor agrees that should there be conflicts or objections not called to the Engineer's attention and written decision rendered by the

Engineer before signing the Contract, the Engineer's decision with regard to such conflict or objections shall be final and binding on the Contractor and shall not be subject to arbitration.

Exhibit 4-20

As can be seen from a quick reading of this clause, the contractor is being asked to assume the risk for any problems that may occur on the job and for any changes that may arise. Clauses of this type do not reduce risk. Instead they generate work for courts, boards, arbitrators, and other dispute forums. It is appropriate at this stage to step back and look at the "big picture" that pulls the contract clauses together.

CONTRACT DRAFTING AND REVIEW

We draft a contract so that we may establish the rules, regulations, procedures, and guidelines under which the construction of the project will be affected. Our goal in this endeavor is twofold. The first goal is to have a contract that is legally sufficient, not in violation of laws or statutes, and that will be enforceable in the courts should the need arise. The second goal is to establish mechanisms in the contract that will enhance the practical execution and management of the project. To achieve these goals, some points of caution should be noted.

To assure that the draft contract is legally sufficient, it is strongly recommended that it be reviewed by an attorney who specializes in construction law. The attorney should have significant experience in handling construction disputes and, therefore, know the common problems to guard against. If you have trouble locating a qualified construction attorney, references are available from the American Bar Association (ABA). The ABA has a Construction Forum Committee and its members generally are specialists in construction law.

The majority of this chapter dealt with practical contract language considerations. Thus, it should help managers achieve the second goal. When drafting contracts, managers should determine their goals, how they want to achieve those goals, what problems they have had in the past, and how they can communicate all of this in the draft contract. It is often beneficial to have the contract reviewed by an outside consultant who has expertise in managing projects. This "extra set of eyes" can provide a more objective and discerning appraisal of the draft.

One final note concerns the manager's review of the draft contract. Always ask yourself if the draft clauses clearly say what you intend. For example, in one project an owner claimed after the fact that it intended to have a broad no damage for delay clause in its contract. The clause read:

> 10.3 *Unavoidable Delays.* Provided Contractor submits to Owner a claim therefor in accordance with subsection 10.2 above, the Contract Time will be extended by the number of days lost due to acts of God, acts of Government, acts of Owner or another contractor in the performance of his contract with Owner, fires, floods, epidemics, quarantine restrictions, strikes and other labor disputes unless caused by jurisdictional disputes arising out of Contractor's use of non-union Subcontractors, freight embargoes, unusually severe weather, and delays of Subcontractors due to such causes. Contractor shall not, however, be entitled to additional compensation nor shall the Contract Price be increased because of any delays arising out of the causes described in this subsection, and Contractor's only remedy under such circumstances will be an extension of the Contract Time.

Exhibit 4-21

Upon first reading one might conclude that this clause allows only extensions of time for any delays that occur. But upon a more careful reading, this is not the case.

First, the clause is titled "unavoidable delays." If it is intended to cover every type of delay then this title seems inappropriate. For instance, if the owner decides to make a change or direct extra work that causes a delay, this is not an "unavoidable" delay. The other specific items in the clause are reasonably unavoidable, such as acts of God, fires, and so on.

Second, the punctuation of the clause is not correct if the intent is to have a broad no damage for delay clause for all acts of the owner. If the intent was to preclude damages for all acts of the owner, then the clause should have had a comma after "acts of owner."

Third, the structure of the clause is not correct, given the presumed intent noted above. Each entity whose acts are being disclaimed is separately identified: "acts of God," "acts of government," and so on. To structure this correctly would require this wording: "acts of government, acts of owner, acts of another contractor. . . ."

As presently written, the clause excludes damages for delay for "acts of Owner or another contractor in the performance of his contract with Owner." In other words, if any delays occur because of work on another contract, whether caused by the owner or the contractor, it is these delays that are noncompensable. This interpretation is a reasonable one. Therefore, if the owner still believes it means all acts of the owner, then the clause is ambiguous. And as a qualified construction attorney will point out, ambiguities are construed against the drafter.

If the owner intended to have a broad no damage for delay clause, the following two examples would have more clearly expressed it.

> 10.3 *Delays.* Provided Contractor submits to Owner a claim therefor in accordance with subsection 10.2 above, the Contract Time will be extended by the number of days lost due to acts of God, acts of Government, acts of Owner, acts of another contractor in the performance of his contract with Owner, fires, floods, epidemics, quarantine restrictions, strikes and other labor disputes unless caused by jurisdictional disputes arising out of Contractor's use of non-union Subcontractors, freight embargoes, unusually severe weather and delays of Subcontractors due to such causes. Contractor shall not however, be entitled to additional compensation nor shall the Contract Price be increased because of any delays arising out of the causes described in this subsection, and Contractor's only remedy under such circumstances will be an extension of the Contract Time.

Exhibit 4-22

> 10.3 *Delays.* Provided Contractor submits to Owner a claim therefor in accordance with subsection 10.2 above, the Contract Time will be extended by the number of days lost due to any causes which are not foreseeable or within the control of the Contractor or its Subcontractors. Contractor shall not, however, be entitled to additional compensation nor shall the Contract Price be increased because of any delays and Contractor's only remedy under such circumstances will be an extension of the Contract Time.

Exhibit 4-23

SUMMARY

The contract's general conditions constitute one of the most important tools to control risk on a construction project. They can set the tone for the entire project. They should be carefully drafted and meticulously reviewed. It is advisable that the drafter of the contract seek qualified help in the drafting of the general conditions. Likewise, it is not advisable to utilize standard contract forms without reviewing them for both practicality and legal sufficiency and making the changes needed to suit your specific project.

CHAPTER 5

MANAGING TIME—ROADMAP TO SUCCESS

While there are many elements to the management of a successful construction project, this chapter focuses on time management, the most critical element of the management process. The core of time management is the proper use of a project schedule. The schedule is far more than what it is often thought to be. Most practitioners believe that the schedule is used only for tracking the progress of the job compared to the required end date. This is a shortsighted view of what a schedule is and what it can do for you.

To manage time, we must first make an accurate estimate of the project duration and the activities that make up the project. Then we must accurately monitor the progress of the project with respect to those time estimates. The process is not simple.

All projects require planning. The better the plan, the more successful the project should be. The entire plan for the project can and should be driven by an overall project schedule, which is a representation of the plan.

It is not uncommon to meet the salty construction manager who will smile when you mention schedules and claim that a formal schedule is unnecessary because they have been building projects just like this one for 40 years. If you suggest that the schedule will help plan the job, they respond that they have a schedule in mind and putting it down on paper is just a needless exercise. It is the intent of this chapter to dispel this erroneous belief and also to provide some techniques that will assist the manager to function effectively.

Unfortunately, the vast majority of construction projects lack a reasonably detailed schedule. There are many reasons for this. The resistance noted above is one reason. A second reason is that a good schedule takes time to prepare and use and we all know that there is never enough time to do everything that we should do. A third reason is that most managers do not fully understand the schedule as a management tool and, therefore, cannot perceive the benefit that it will bring to the success of the project.

When schedules are used, they generally are not implemented correctly or are resorted to only after problems have arisen related to time. On a large airport project, the aviation authority wanted an independent evaluation of the reasonableness of the construction manager's request for additional funds to accelerate the project to make up a project delay of seven weeks. Since a CPM schedule was required by the contract, it was one of the first items the authority requested to review. The schedule could not be located for a full two days. Obviously, it had not been used as a management tool during the course of the job.

Commonly, we see schedules that apply to the construction phase of the project. While this is a good step, the schedule should really be established to control the entire project. It should begin at the inception of the project and continue until the very end. It can and should include the many tasks that the owner must perform such as the feasibility studies, owner decisions and reviews, the entire design process, approval meetings, owner procurement, bid issue and award, and finally the actual project construction with clearly defined milestones.

In general, projects are run with one of two types of schedules, either a bar, or Gantt, chart or a CPM schedule. A third form of schedule that is sometimes used is called a linear schedule. The linear schedule, however, is very limited in its application and effectiveness and, therefore, will not be addressed in this book.

BAR CHARTS

Henry Gantt developed bar charts in the early twentieth century. Bar charts are the most common form of schedule used on a construction project because they are simple to use and take minimum effort to develop and update. There are several advantages and disadvantages to the use of bar charts. These can be summarized as:

Advantages

Easily understood by all parties

Easy to draft an original schedule
Easy to update the schedule

Disadvantages

Schedule lacks detail
Relationship of activities is not shown
Difficult to determine project status

Virtually anyone can understand a bar chart. This makes it an ideal tool to communicate a complex plan in a simple fashion. However, if the plan is complex, the simplification provided by the bar chart may dilute the plan to the extent that the chart's use as a tool to control and monitor the work is completely subverted. For reasons of simplicity and also because of lack of effort, most bar charts have far too few activities to accurately describe the work to be performed. For example, a typical bar chart for the construction of a four-span highway bridge is shown in Figure 5-1.

Figure 5-2 shows a simple elevation view of the bridge. The bridge has three piers on pile foundations, abutments, approaches, steel spans, and a concrete deck and curbs.

The bar chart does not distinguish when each pier will be constructed, nor does it detail when the spans will be built. Consequently, the lack of detail makes it very difficult to determine exactly how the job is planned or will be executed with respect to time. If the plan is not clear in the first place, the process of monitoring the work becomes more guesswork than anything else. Furthermore, the schedule does not show how the activities are related to one another. This is an extremely serious shortcoming since it is not possible to determine which activities are critical to the completion of the project.

Since much of the discussion in this chapter will use the term "critical," it should be defined at this point. An activity is critical if a delay to that activity will delay the final completion date of the project. Whether a CPM schedule is used, there are still critical and noncritical activities on every project.

Most bar charts are limited to less than 40 activities. The practical reason for this is that 40 activities are the most that can easily fit on a single sheet of paper. The size of the paper dictates how many activities are included in the bar chart.

The bar chart can be made more useful as a management tool by increasing the level of detail used. For example, taking the bar chart for our sample project from Figure 5-2 and refining the activities will result in the bar chart shown in Figure 5-3.

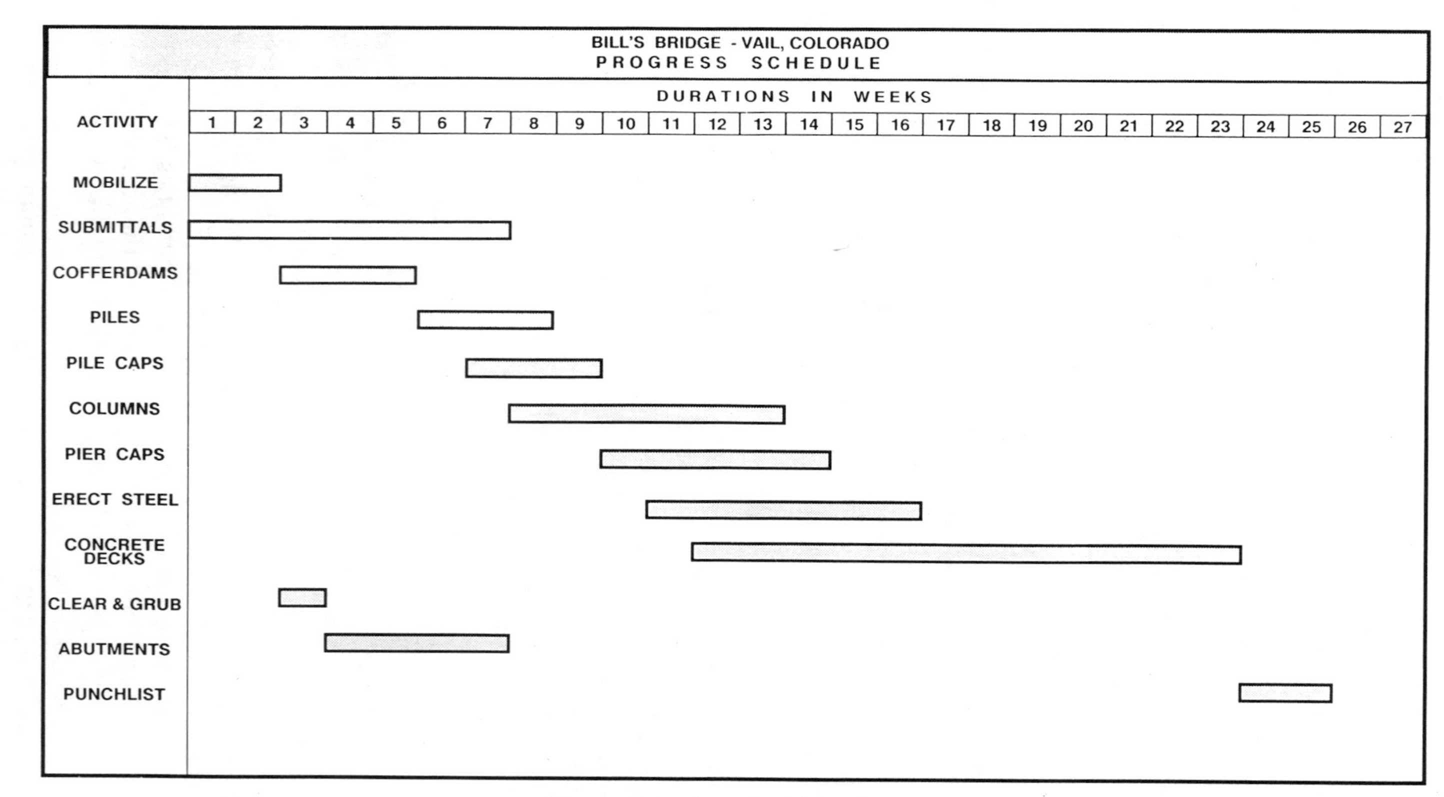

Figure 5-1. A typical bar chart.

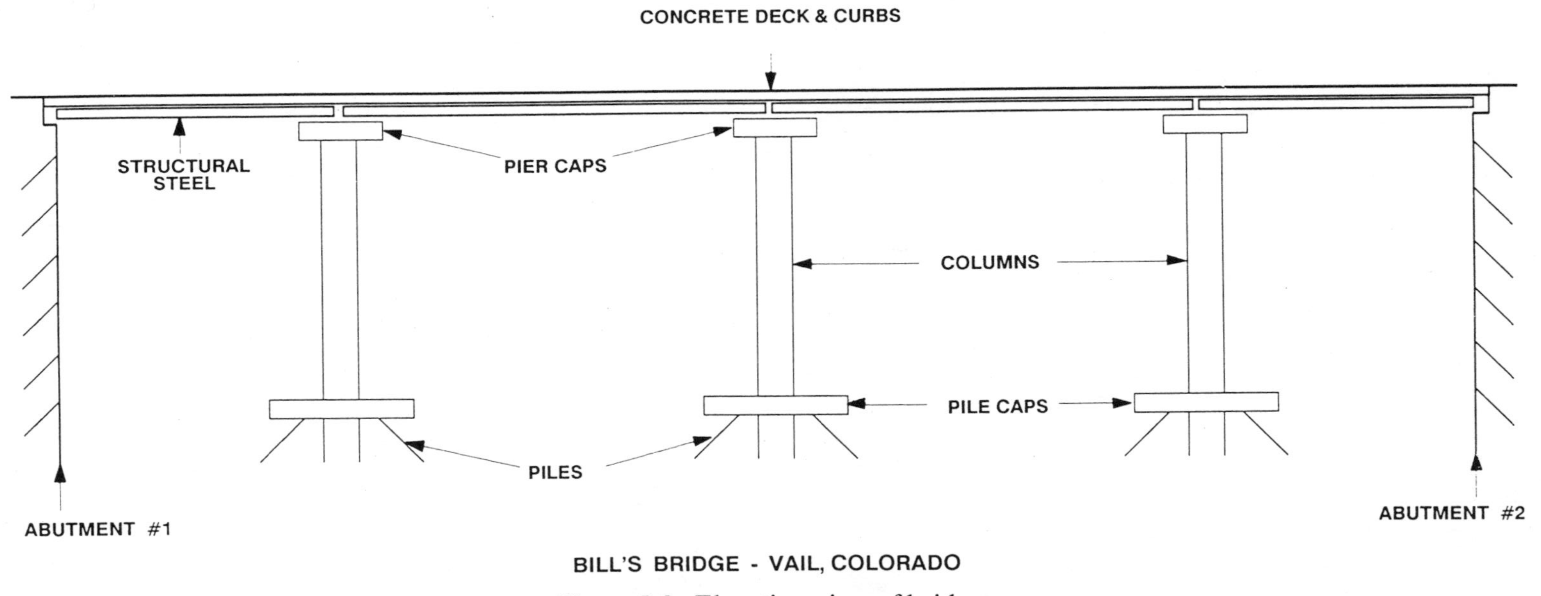

Figure 5-2. Elevation view of bridge.

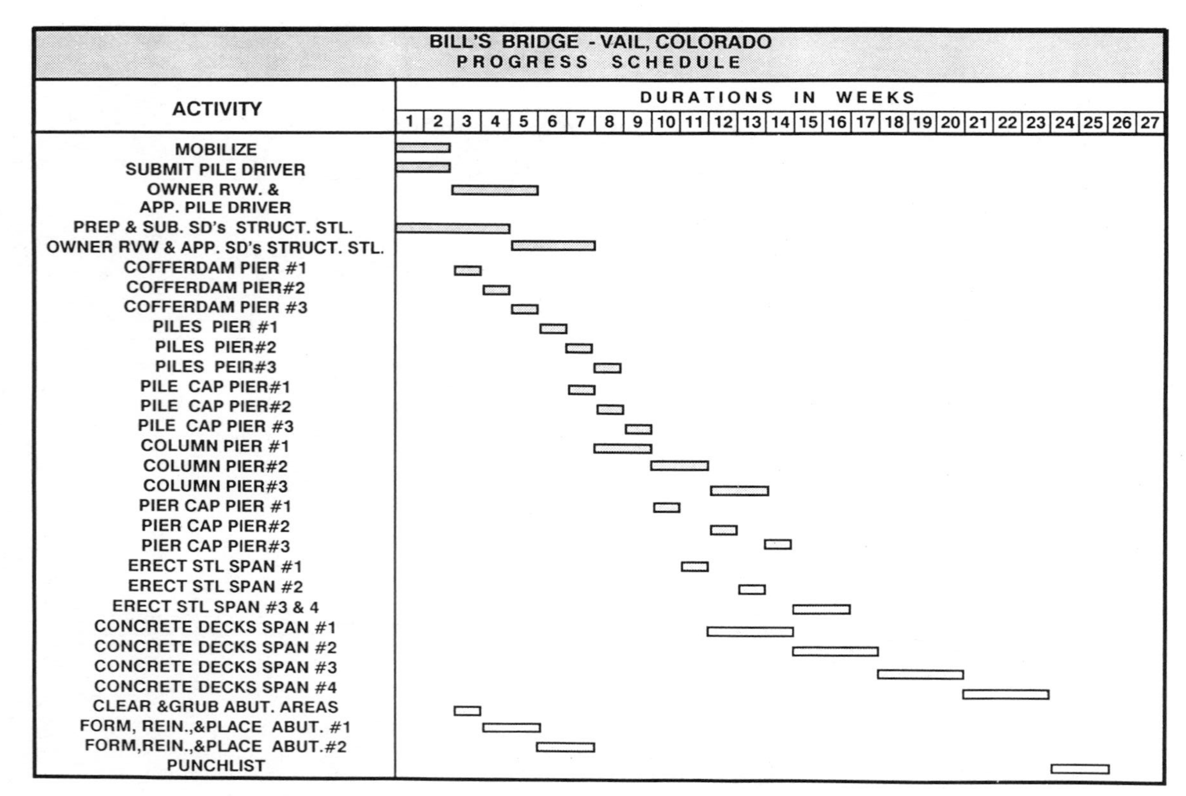

Figure 5-3. Original chart refined.

As Figure 5-3 shows, the piles, pile caps, piers, and pier caps are now clearly identified in the schedule. A similar definition is provided for the steel for each span, the approaches, and the concrete decks and curbs. Obviously, this more detailed schedule will better serve as a management tool. It should be noted that the effort involved in refining the schedule was not significant.

Historically, bar charts have been used in conjunction with progress curves, often referred to as *S-curves*. A typical S-curve for a theoretical project is shown in Figure 5-4.

The S-curve for the perfect project shows that the project starts at a relatively slow pace, picks up momentum which is represented by the almost straight line portion of the curve, and slows down toward the end of the project. In the perfect project, the initial mobilization and start-up encompasses the first 25% of the time and accomplishes 15% of the work. By the 75% time mark, 85% of the work is performed, and the remaining 15% of work is accomplished during the last 25% of the time. Few projects, however, are perfect.

In plotting an S-curve, units must be established for the horizontal and vertical scales. Typically, the horizontal scale is expressed in time. The vertical scale is expressed either in dollars or in hours. A bar chart is usually established for a project with dollar or hour values assigned to each activity. These then are summed over time to yield the S-curve. Using our sample project and its bar chart from Figure 5-1, we can assign dollar values to the activities and plot our S-curve. This is shown in Figure 5-5.

Some owners require the use of S-curves as a means of measuring progress on a project. The contractor is required to submit an initial S-curve based on a bar chart for the project. Each month, the contractor then overlays the actual "progress" against the planned S-curve and reports a percentage completion. An example of this is shown in Figure 5-6.

This method of tracking progress can be very misleading. It presumes a direct relationship between dollars and time. It further presumes that the dollar values assigned to the activities are correct. Many owners and contractors have been deceived by the use of an S-curve as a means of tracking or reporting progress. Most S-curve progress measurements show the project ahead of schedule during the first half of the job. What this really means is that the contractor has invoiced more than was noted in the dollar breakdown for the activities. There are many reasons why this can occur. For example, front-end loading

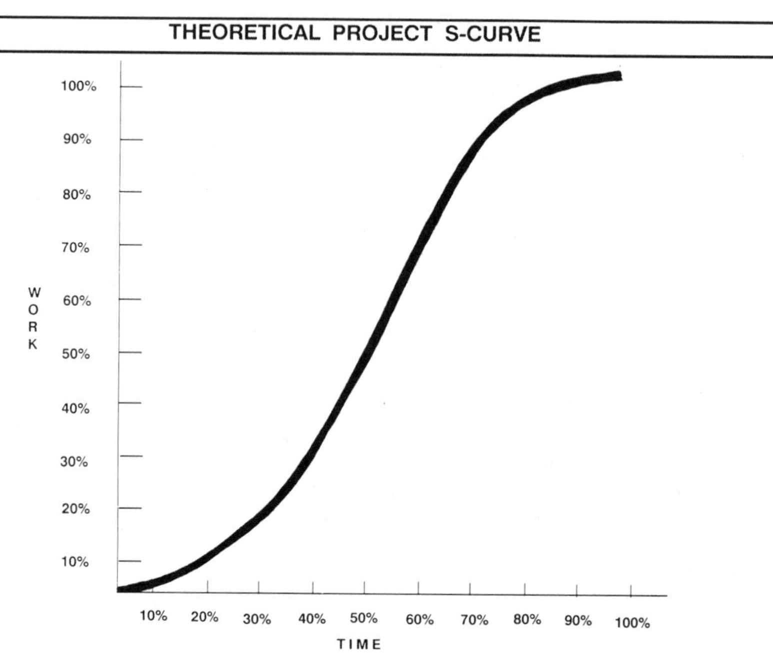

Figure 5-4. S-curve.

of a schedule of values or of pay requests could account for this "ahead of schedule" reporting. It is not necessarily true that the project is ahead of schedule.

While S-curves have some use in the industry, they are not an accurate method for monitoring progress of the job with respect to time. More appropriate uses for S-curve will be noted later in this chapter.

The misconception of a linear relationship between dollars and time can cause other problems in the management of a construction project. The major problem occurs with time extensions for extra work. Some construction contracts specify that time extensions will be granted based on the relative dollar value of a change to the overall value of the project. An example of this type of clause was given in Chapter 4.

Even a cursory analysis of the logic of this would show it to be a fallacious approach to assessing additional time for a project. For instance, an owner may make a change in the type of steel to be used for our sample bridge project. The new steel is more expensive than what was originally specified. Consequently there is a change order for additional cost. Presuming the cost of the new steel is an increase of 10% to the contract and that the original contract time was 200 calendar days, then a time extension of 20 days should be granted. If the steel change is directed early enough and there is no additional time for procurement, in reality then, no time extension is warranted. Therefore, caution must be exercised in making assumptions concerning relationships of time with any other elements, particularly dollars.

In summary, bar charts are simple tools that can be used to help manage the project. Their usefulness is dependent on the level of detail of the bar chart and the manner in which the bar chart is used. Appropriate uses for bar charts will be addressed later in this chapter.

CPM SCHEDULES

Critical path method schedules are the second type of schedule that are used on construction projects. CPMs came into being in the mid-1950s and have been growing in popularity ever since.

Often, CPM is also referred to as *PERT*, the *program evaluation and review technique*. Though similar in some respects, the two are not the same. PERT is primarily used for research and development projects, not construction jobs. The network diagrams for both CPM and PERT look the same, and this is one reason why the terms have been used interchangeably. The major difference between CPM and PERT is that PERT uses probabilistic durations or dates for the durations of activities since they cannot accurately be estimated based on past experience. For our purposes, we will deal strictly with CPM since that is what is used in construction.

It should be noted that this book is not a text on CPM. Our discussion will deal with the application of CPM and how it can be used to manage a project and how it can benefit a job if properly applied. This then discusses only a few of the basics of CPM in order to set the stage for the remainder of the discussion.

Though CPM sounds intimidating to the first-time user, it is a very simple process. A CPM is an arrow or network diagram which graphically represents how one intends to build a project. You may wonder why, if the process is so simple, CPM isn't used on all projects.

Many reasons are given for not using a CPM, including:

It is too complex for my job.
It is too theoretical.
It is too expensive.
It takes too much time to do.

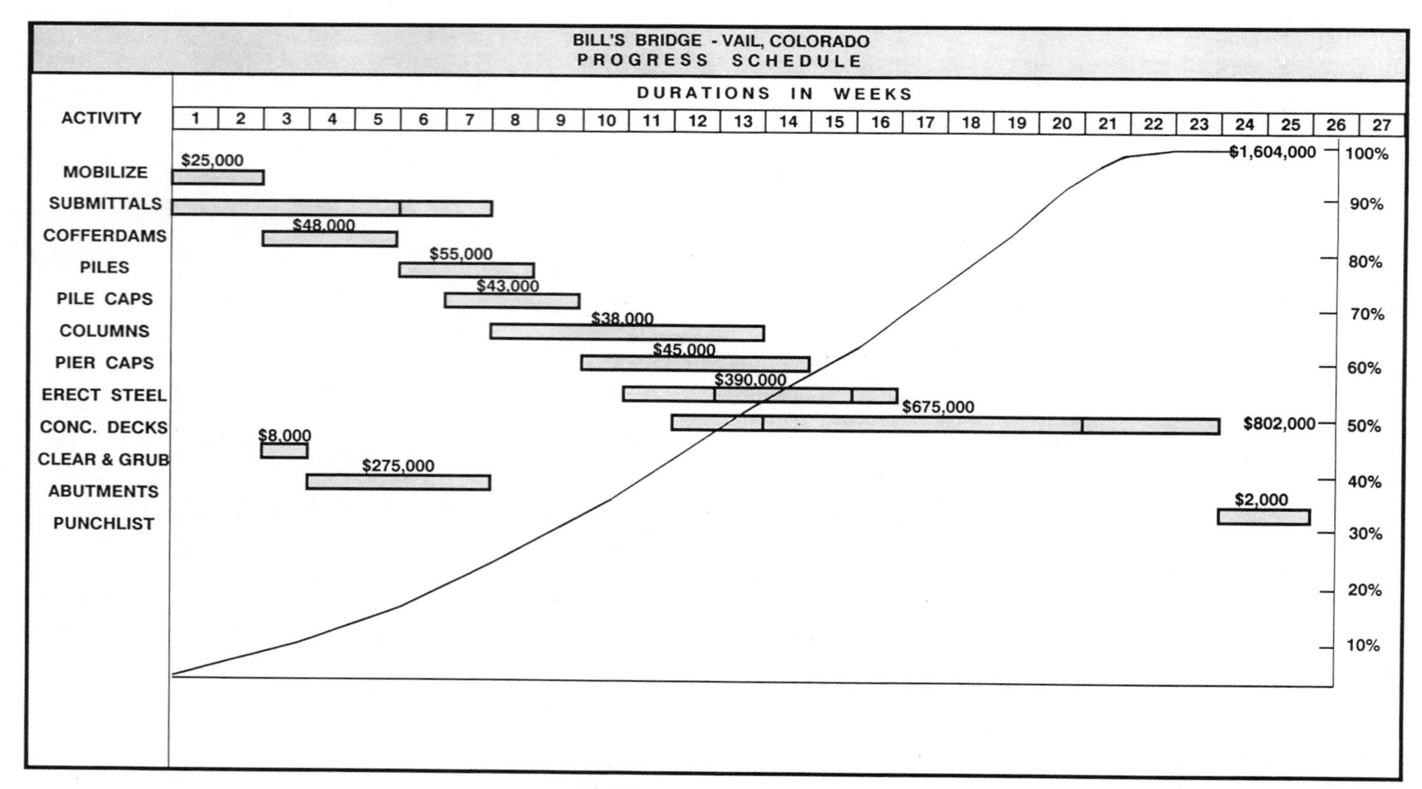

Figure 5-5. Generating the S-curve.

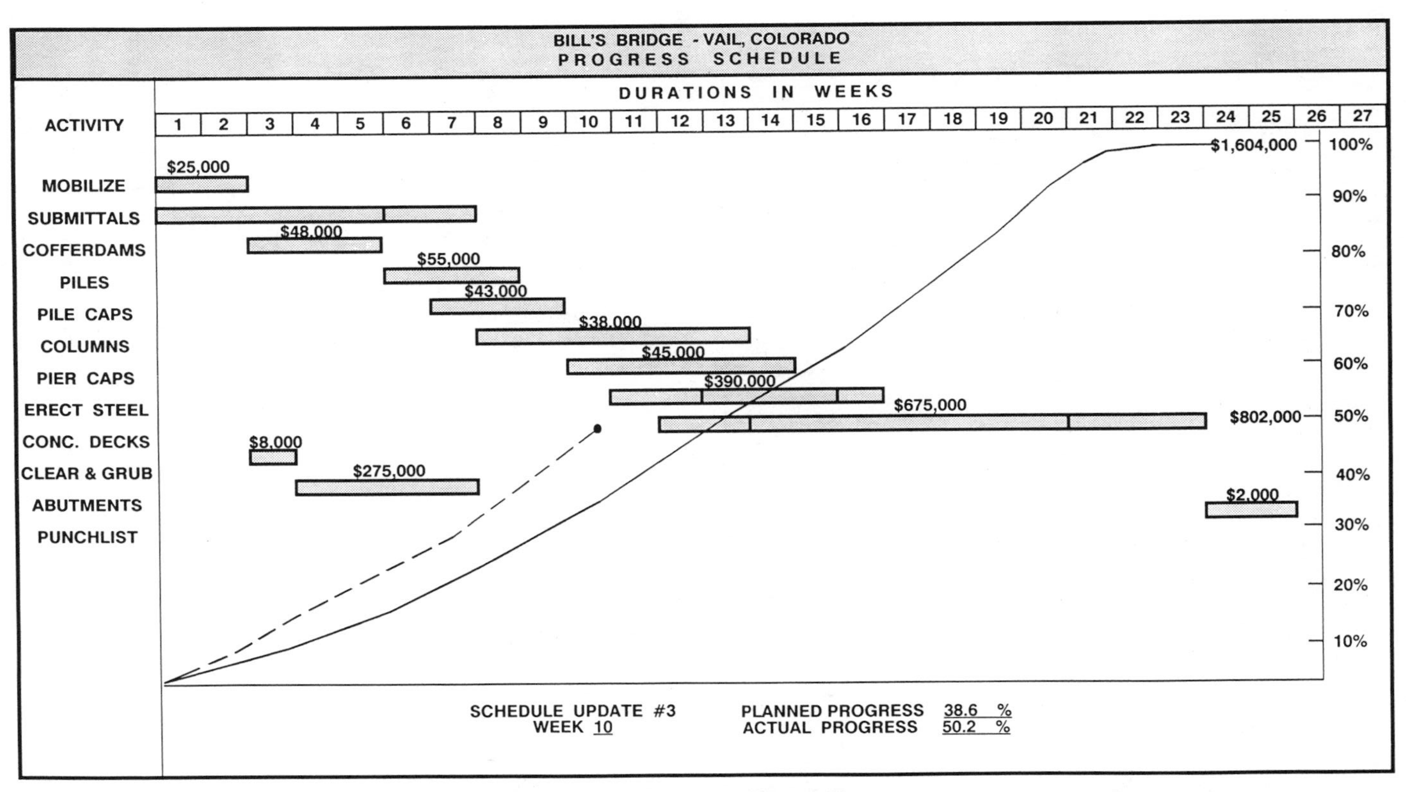

Figure 5-6. Estimated vs. "actual" progress.

These excuses are without foundation. In reality, the reason a CPM is not used is because it is not understood or it is not correctly applied.

The CPM is an extremely powerful tool that can aid in planning for manpower, equipment, and cost, as well as time. There are two general types of CPM schedules: *activity-on-arrow* and *activity-on-node*, or *precedence.*

Activity-on-Arrow Diagrams

An activity-on-arrow schedule is simply an arrow diagram showing the logic or interrelationship among various activities. An example of this type of diagram is shown in Figure 5-7. As shown in Figure 5-7, the arrows represent specific activities with defined durations. The circles or nodes are events or points in time with no duration. This form of CPM was the type originally developed in the mid-1950s.

Activity-on-Node Diagrams

The activity-on-node, or precedence, CPM is also a simple diagram showing the activities, their interrelationships, and durations. An example of this is shown in Figure 5-8. As noted in Figure 5-8, the boxes or nodes represent specific activities with defined durations. The arrows represent events or points in time with no duration. In simplistic terms, one form is almost the opposite of the other. Why then have two different formats?

With an activity-on-arrow CPM, interrelationships may have to be shown by the use of activities with zero duration, which are called "dummies" or "dummy activities." These are also shown in Figure 5-7. The dotted line joining activity 15–20 is showing that activity 10–15 must be completed before activity 20–35 can be started. The use of dummies adds additional activities to the overall schedule.

In precedence schedules, no dummies are used. Instead, the scheduler can control the interrelationships by the use of *lead/lag* factors or relationships. This not only eliminates dummy activities but can also be used to eliminate actual activities. For example, Figure 5-9 is a precedence diagram with one of the lead/lag factors noted. The specific relationship shown is a lag of three days between the finish of activity 20 and the start of activity 35. Activity 20 is the placing of concrete while activity 35 is the stripping of forms. The three-day lag is replacing an activity for the required cure time of three days. The lag relationship is saying that activity 35 cannot start until three days after the finish of

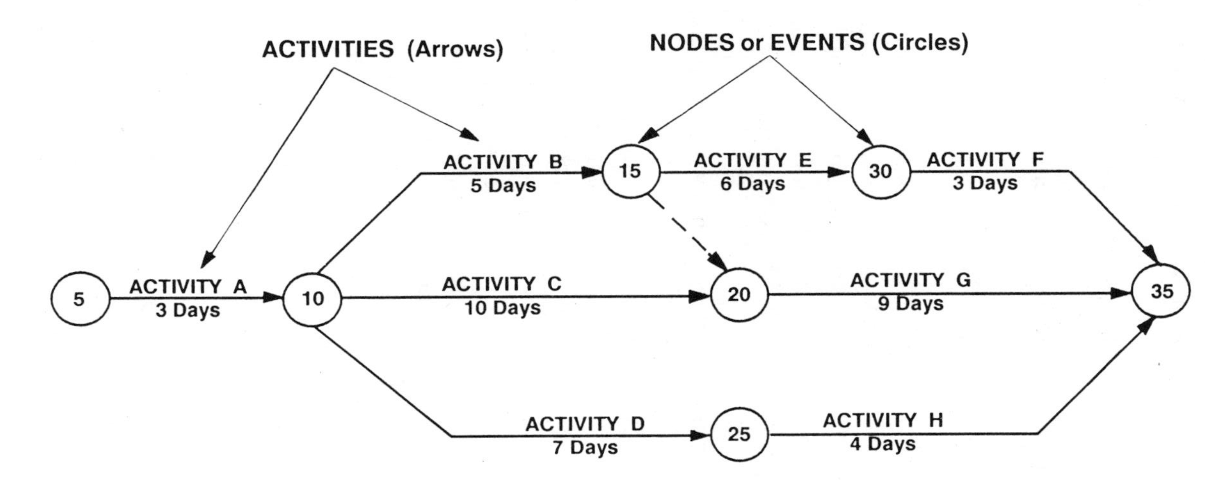

Figure 5-7. Sample activity-on-arrow CPM schedule.

activity 20. Obviously, the use of lead/lag factors can reduce the overall size of the schedule and the diagram. In the most technical sense, precedence schedules are more powerful and efficient to use than activity-on-arrow schedules. Unfortunately, however, the manager must be extremely proficient in CPM scheduling to effectively use a precedence schedule. Similarly, the other parties who might be using the schedule must also be proficient in their use. Because of the general lack of proficiency in the industry, the author recommends the use of activity-on-arrow schedules instead of precedence schedules.

Constructing a CPM Schedule

In constructing a CPM schedule, the manager should be aware of several guidelines that will make the schedule more effective. First, the activity descriptions should be clear to avoid any misunderstanding of what they mean. For instance, an activity labeled "drainage" could mean site drainage, scuppers on a bridge, or any number of other types of drainage.

Second, the schedule must be constructed with a reasonable level of detail. This level of detail is dependent on the use of the schedule. For instance, during the early planning stages, one might create a milestone form of schedule with activity durations of four to six months. The purpose of the schedule is solely to lay out the "big picture" and not to plan and control the project during the day-to-day operations. Likewise, the detail of the schedule will depend on the nature of the project. The simpler the project is, the less detail will be required.

In setting up the detailed construction schedule, a general guideline is to use activities with durations not less than 1 day nor more than 15 days. Except for unique projects, using durations of less than 1 day tends to create too many activities. Since the schedule is a tool to help monitor the progress of the work, a manager would not practically monitor activities of less than a day. Taken to the extreme, we would be on the project with stopwatches. This is not the intent of project scheduling. Conversely, using activities with durations greater than 15 days makes the job of assessing progress on a specific activity too subjective. For instance, if you were reviewing the physical progress on a project and looked at an activity with a duration of 120 days, your assessment of the amount of work or percentage of the activity completed would be far less accurate than if you were reviewing an activity with a duration of 15 days. Shorter durations also preclude the various contractors from "dancing" when asked about their respective progress. When quizzing a subcontractor, many a general contractor has

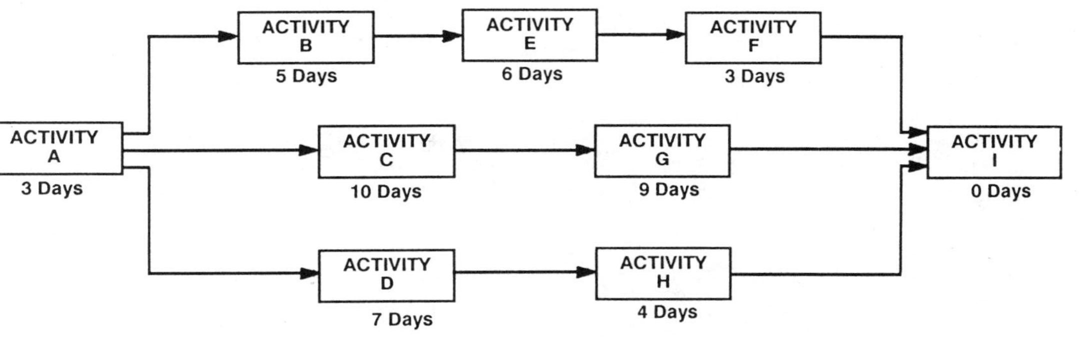

Figure 5-8. Sample activity-on-node CPM schedule.

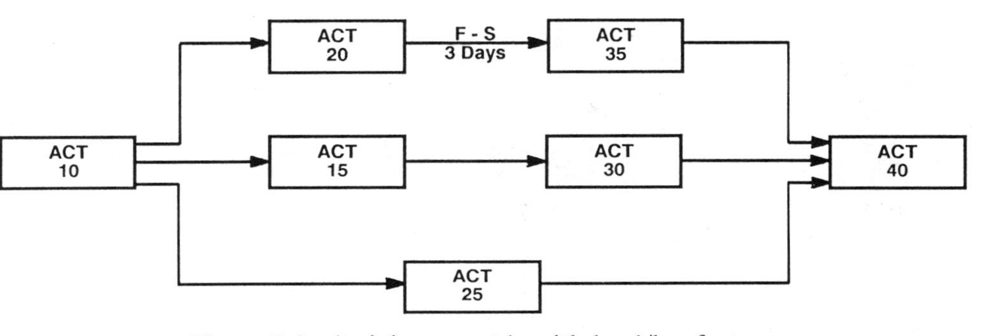

Figure 5-9. Activity-on-node with lead/lag factor.

heard that the subcontractor's crews always start a little slowly but pick up speed as the work progresses. If this is true, then the shorter duration activities will reflect the commensurate degree of work planned and the manager can make a more reasoned assessment of the actual progress of the work.

In scheduling projects, one problem is determining how to break up an activity with a duration that should obviously be longer than 15 days. For example, assume the activity is installing drywall over a large floor area of a building. The activity will encompass 45 days based on the crew size planned. In this case, the activity can be broken down into three parts, based on physical areas on the floor such as from column line A to D, from column line D to G, and from column line G through I. If the subcontractor is planning for a slow start and more rapid performance as it moves across the floor, than the activities might have durations of 17 days, 14 days, and 14 days, respectively.

In assigning durations to the respective activities, careful consideration must be given to an assessment of the resources that will be used, including manpower, equipment, number of shifts, and material. Ignoring any of these considerations can result in assigning incorrect or unattainable durations. For example, in preparing a CPM schedule, a general contractor planned for concrete paving operations that were extremely aggressive. When the project was underway, the contractor had the manpower and the equipment to meet this aggressive schedule, but the supplier of concrete material could not keep up with the planned paving production. As a consequence, the contractor was forced to idle some of his equipment and took longer to complete the work than originally planned. Consider *all* resources.

When assigning durations, incorporate the advice and decisions of the parties who will actually be performing the work. If subcontractors are involved, include them in the scheduling process. If they establish durations for their respective activities, they should be more accurate and there will be fewer problems during the actual execution of the work. If you impose your durations on another party, they might later assert that you directed their method and manner of performing the work. As a consequence, they could claim that since you usurped their ability to manage their work, any costs in excess of their estimate are your liability. Also, you might receive less cooperation in achieving your estimated durations.

At this stage, you may be saying that this all sounds "too complex." This section began with the premise that CPM schedules are simple. Let's prove that point. Refer back to Figure 5-3, the bridge job. We refined the activities from our first pass in the Figure 5-1 schedule. If we now take these refined activities and reflect on why we drew our bar chart in the manner shown, we can make a few easy modifications and be very close to our CPM schedule. The activities for pile driving for the three piers are shown to occur sequentially. We made this decision because we desired to use one pile driving crew and move them from pier to pier. As soon as the piles are driven for pier 1, we mobilize a separate crew to construct the pile caps. This crew moves through the piers in the same fashion as the pile driving crew. Once a pile cap is constructed, we planned for a separate crew to begin the pier columns. Based on this plan, we can show on our bar charts how the activities interrelate. This is shown in Figure 5-10.

The arrows drawn on our simple bar chart now show how the flow of work is planned for the job. If we now do a bit of minor redrafting, we have Figure 5-11, an activity-on-arrow diagram and our project CPM. The process of refining the detailed bar chart into a rough network diagram is not difficult. It is primarily a function of how you plan the project. For example, in our sample project, the manager has decided to use one crew for construction of all three cofferdams. Because of this, an arrow is drawn between the end of cofferdam pier #1 and the start of cofferdam pier #2. A similar relationship is shown between the end of cofferdam pier #2 and the start of cofferdam pier #3. In a similar fashion, the project manager has decided to utilize only one crane for the work on the water. Because of this, the schedule shows with an arrow from the end of cofferdam pier #3 to the start of piles pier #1 that the pile driving starts after all the cofferdams are completed. The remaining relationships are determined by similar resource considerations, by the physical nature of the project, or even by arbitrary choice. An example of the physical nature of the project would be the interrelationship shown between the finish piles pier #1 and the start of pile cap pier #1. As you can see, it really is simple.

When a project schedule is being developed, the scheduler must be careful not to let the contract specified duration control the durations assigned to each activity. For instance, many project managers, when laying out a schedule, set up their logic, and then fix durations so that the schedule "fits" the contract time. It is more correct to assess each activity based on the amount of work required and the resources that will be applied, and determine the duration based on that. Once all durations are established, the managers could see what the overall duration of the project is. If the duration is too long, the managers must either resequence activities, apply more resources, or work faster than planned. This then serves as an early warning that a project requires more than just an ordinary effort to finish on time. Conversely, it might turn out that the overall duration is shorter than the contract specified time. In this case, the manager can plan and work for an early completion.

Early Completion Schedule. There is nothing inherently wrong with an early completion schedule. In fact, in today's highly competitive market, successful contractor firms are those who develop methods to perform the work in less time, thereby reducing project overhead and costs and offering more competitive bids. If the owner receives a schedule showing less time than allowed in the contract, it should not reject the schedule for not complying with the contract. The contract normally only specifies that the project must be completed within a certain number of days. It does not say that it cannot be finished earlier. In the face of an early completion schedule, the owner should review the schedule carefully to insure there are no major problems. If necessary, additional discussions should be held with the contractor regarding the schedule to be sure the owner has no doubts. Then, during the course of the project, the owner should carefully monitor progress within the context of the schedule to insure that time goals are being met. If they are not, the reasons for slippages should be well documented to avoid a claim for delay to early completion of the project.

Many contractors develop a project schedule that shows they can complete early. This schedule, however, is not submitted to the owner. Instead a different schedule for the entire contract duration is submitted and the contractor works from the early completion schedule. This is not an advisable practice; it can only lead to problems. First, the owner needs to know if the contractor is planning on an early completion since it will affect the owner's cash flow for paying the contractor's invoices. Second, if a delay caused by the owner occurs, the contractor may be hard pressed to show why it was delayed and explain why it had two different schedules on the job. Finally, it is almost certain that a subcontractor firm will see the longer schedule posted somewhere on the project, complain to the contractor that it is being accelerated because of the early completion schedule, and perhaps even refuse to work toward the shorter time frame.

CPM as a Management Tool

While the CPM process defines those activities that are critical or cannot be delayed without delaying the overall completion date of the project, the schedule will also show those activities that are not critical. Those activities are said to have *float*, or *slack*, time.

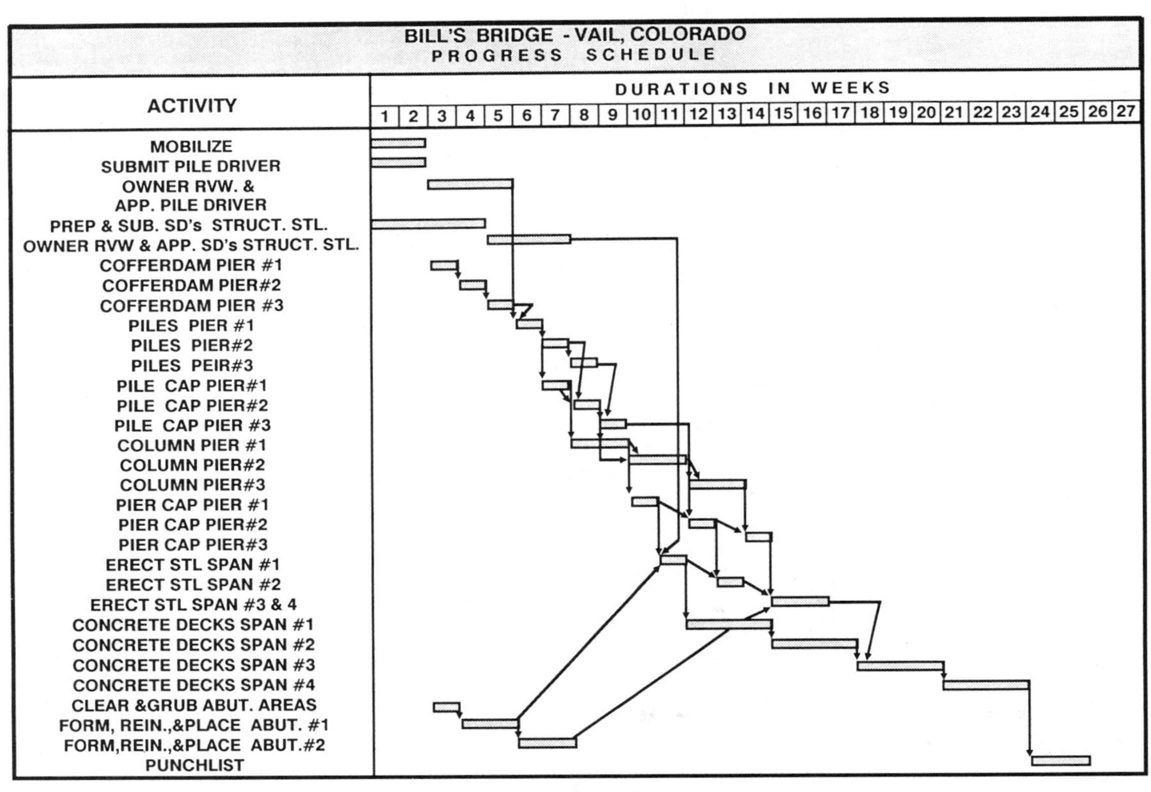

Figure 5-10. Interrelationships shown more precisely.

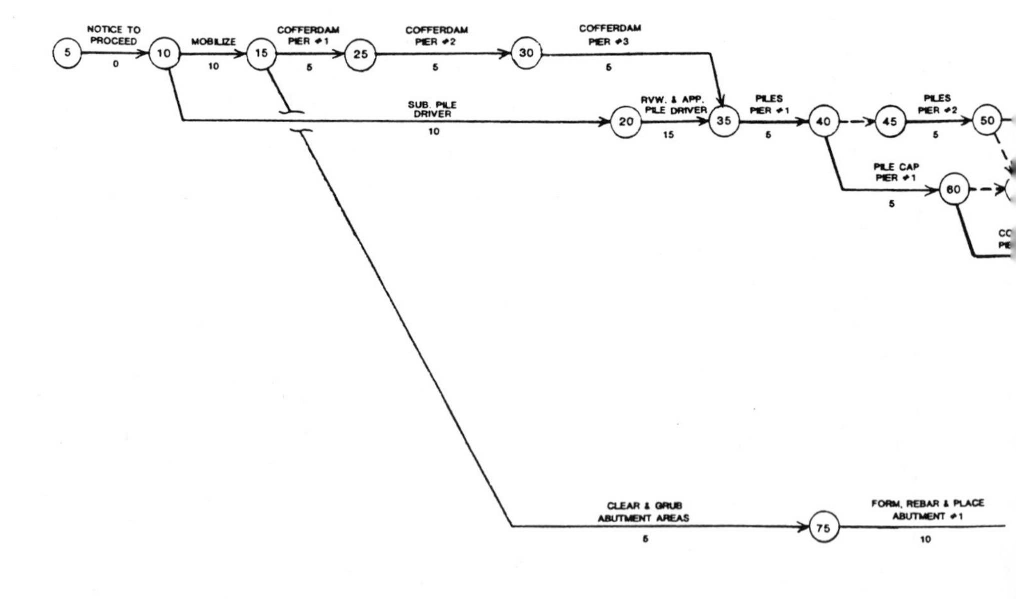

Figure 5-11. Schedule's precision depicted on activity-on-arrow diagram.

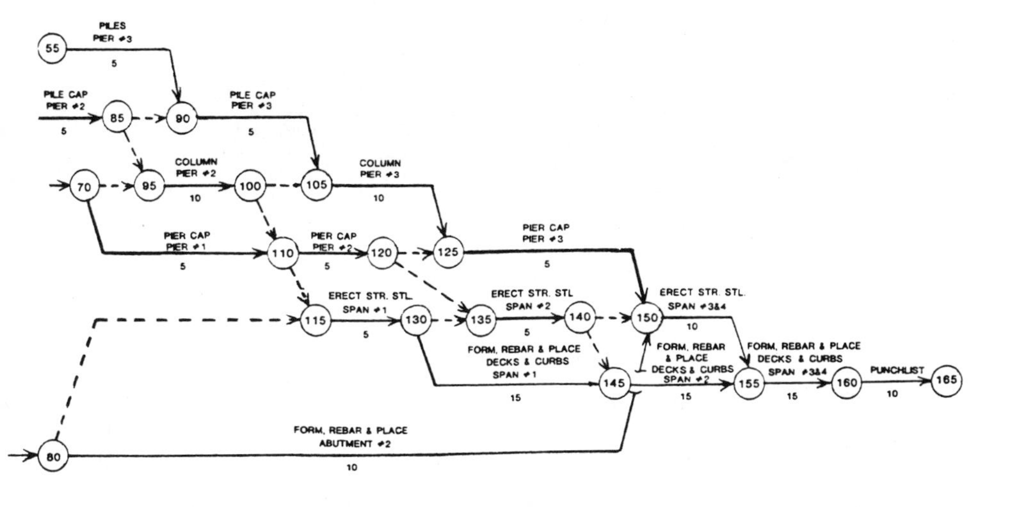

Figure 5-11. *(Continued)*

TRAUNER CONSULTING SERVICES, INC. PRIMAVERA PROJECT PLANNER

REPORT DATE 3APR92 RUN NO. 19 START DATE 3MAY93 FIN DATE 8OCT93
10:45
SCHEDULE REPORT - SORT BY TF, ES DATA DATE 3MAY93 PAGE NO. 1

PRED	SUCC	ORIG DUR	REM DUR	%	CODE	ACTIVITY DESCRIPTION	EARLY START	EARLY FINISH	LATE START	LATE FINISH	TOTAL FLOAT
5	10	0	0	0		NOTICE TO PROCEED	3MAY93	30APR93	3MAY93	30APR93	0
10	15	10	10	0		MOBILIZE	3MAY93	14MAY93	3MAY93	14MAY93	0
10	20	10	10	0		SUB. PILE DRIVER	3MAY93	14MAY93	3MAY93	14MAY93	0
15	25	5	5	0		COFFERDAM PIER #1	17MAY93	21MAY93	17MAY93	21MAY93	0
20	35	15	15	0		RVW & APP. PILE DRIVER	17MAY93	4JUN93	17MAY93	4JUN93	0
25	30	5	5	0		COFFERDAM PIER #2	24MAY93	28MAY93	24MAY93	28MAY93	0
30	35	5	5	0		COFFERDAM PIER #3	31MAY93	4JUN93	31MAY93	4JUN93	0
35	40	5	5	0		PILES PIER #1	7JUN93	11JUN93	7JUN93	11JUN93	0
40	60	5	5	0		PILE CAP PIER #1	14JUN93	18JUN93	14JUN93	18JUN93	0
60	70	10	10	0		COLUMN PIER #1	21JUN93	2JUL93	21JUN93	2JUL93	0
70	95	0	0	0		DUMMY	5JUL93	2JUL93	5JUL93	2JUL93	0
95	100	10	10	0		COLUMN PIER #2	5JUL93	16JUL93	5JUL93	16JUL93	0
100	110	0	0	0		DUMMY	19JUL93	16JUL93	19JUL93	16JUL93	0
110	115	0	0	0		DUMMY	19JUL93	16JUL93	19JUL93	16JUL93	0
115	130	5	5	0		ERECT STRUCTURE STEEL SPAN #1	19JUL93	23JUL93	19JUL93	23JUL93	0
130	145	15	15	0		FORM,REBAR, AND POUR DECKS & CURBS SPAN #1	26JUL93	13AUG93	26JUL93	13AUG93	0
145	155	15	15	0		FORM,REBAR, AND POUR DECKS & CURBS SPAN #2	16AUG93	3SEP93	16AUG93	3SEP93	0
155	160	15	15	0		FORM,REBAR, AND POUR DECKS & CURBS SPAN #3 & #4	6SEP93	24SEP93	6SEP93	24SEP93	0
160	165	10	10	0		PUNCH LIST	27SEP93	8OCT93	27SEP93	8OCT93	0
70	110	5	5	0		PIER CAP PIER #1	5JUL93	9JUL93	12JUL93	16JUL93	5
100	105	0	0	0		DUMMY	19JUL93	16JUL93	2AUG93	30JUL93	10
105	125	10	10	0		COLUMN PIER #3	19JUL93	30JUL93	2AUG93	13AUG93	10
110	120	5	5	0		PIER CAP PIER #2	19JUL93	23JUL93	2AUG93	6AUG93	10
120	135	0	0	0		DUMMY	26JUL93	23JUL93	9AUG93	6AUG93	10
130	135	0	0	0		DUMMY	26JUL93	23JUL93	9AUG93	6AUG93	10
135	140	5	5	0		ERECT STRUCTURE STEEL SPAN #2	26JUL93	30JUL93	9AUG93	13AUG93	10
125	150	5	5	0		PIER CAP PIER #3	2AUG93	6AUG93	16AUG93	20AUG93	10
140	145	0	0	0		DUMMY	2AUG93	30JUL93	16AUG93	13AUG93	10
150	155	10	10	0		ERECT STRUCTURE STEEL SPAN #3 & #4	9AUG93	20AUG93	23AUG93	3SEP93	10
120	125	0	0	0		DUMMY	26JUL93	23JUL93	16AUG93	13AUG93	15
140	150	0	0	0		DUMMY	2AUG93	30JUL93	23AUG93	20AUG93	15
40	45	0	0	0		DUMMY	14JUN93	11JUN93	12JUL93	9JUL93	20
45	50	5	5	0		PILES PIER #2	14JUN93	18JUN93	12JUL93	16JUL93	20
50	55	0	0	0		DUMMY	21JUN93	18JUN93	19JUL93	16JUL93	20
50	65	0	0	0		DUMMY	21JUN93	18JUN93	19JUL93	16JUL93	20
55	90	5	5	0		PILES PIER #3	21JUN93	25JUN93	19JUL93	23JUL93	20
60	65	0	0	0		DUMMY	21JUN93	18JUN93	19JUL93	16JUL93	20
65	85	5	5	0		PILE CAP PIER #2	21JUN93	25JUN93	19JUL93	23JUL93	20
85	90	0	0	0		DUMMY	28JUN93	25JUN93	26JUL93	23JUL93	20
90	105	5	5	0		PILE CAP PIER #3	28JUN93	2JUL93	26JUL93	30JUL93	20
15	75	5	5	0		CLEAR & GRUB ABUTMENT AREAS	17MAY93	21MAY93	28JUN93	2JUL93	30
75	80	10	10	0		FORM,REBAR, AND POUR ABUTMENT #1	24MAY93	4JUN93	5JUL93	16JUL93	30
80	115	0	0	0		DUMMY	7JUN93	4JUN93	19JUL93	16JUL93	30
80	150	10	10	0		FORM,REBAR, AND POUR ABUTMENT #2	7JUN93	18JUN93	9AUG93	20AUG93	45

Figure 5-12. Float vs. critical path.

Float. *Float* is defined as the difference from the time an activity can start and the time it must start. For instance, we can take our sample bridge project and perform the arithmetical calculations required by the CPM method. The results of this would show the critical path and the activities with float. The results of this calculation are shown in Figure 5-12. As can be seen from the total float sort shown in Figure 5-12, activity 110–120 has float of 10 work days or 14 calendar days. The calendar day float can be seen from the difference in the calendar dates between the early start and the late start. This means that the activity can start as early as July 19, 1993, or can start as late as August 2, 1993, and still not delay the overall duration of the project. This definition of float is technically called *total float*.

A second type or definition of float involves what is known as *free*, or *noninterfering*, float. This is float which, if used, will not delay the early start of a succeeding activity. For example, Figure 5-12 shows activity 150–155 has 10 days of float. The succeeding activity is activity 155–160, with an early start of September 6,1993. If activity 150–155 starts three days later than its early start date, it will have no effect on the succeeding activity 155–160. The subtle distinction between total float and free float is important to the manager. In fact, the entire concept of float is important.

Many managers have the mistaken belief that float is time that can be "wasted." In reality, float is time that should be used to increase profits. In order to use float to your advantage, you must understand what it is, how it affects the schedule and the respective activities, and what elements can benefit from the proper use of float. In general, manpower and equipment are the items that can be optimized by the proper use of float.

Resource-Loaded Schedules. When we developed our project schedule, we first defined the logic of how we were going to build the job. Once we refined the bar chart into a logic diagram, we assigned durations to activities based on all of the resources required. When we were considering the resources, we could very easily have noted on our diagram the amount of resources that we had determined for each activity. For example, referring back to our logic diagram on our sample project, we can note the resources as shown in Figure 5-13.

As can be seen in Figure 5-13, activity 15–25 has been assigned a crew of four operators and six laborers. Similarly, activity 35–40 has been assigned one pile driver, one crane, and one barge. By doing this for all activities, you have constructed a *resource-loaded CPM schedule*. There are many benefits to the use of such a schedule. First, it virtually eliminates the argument that a contractor is not applying the correct amount of effort. Far too many projects have a running debate over whether or not the contractor has enough workers on the job. Since the schedule has already defined the number of workers by trade, either they are there or they aren't. Second, the use of assigned resources allows an objective check of the logic and durations of the schedule. For example, a reviewer can look at a specific activity, determine the work involved, and see if the appropriate amount of resources has been

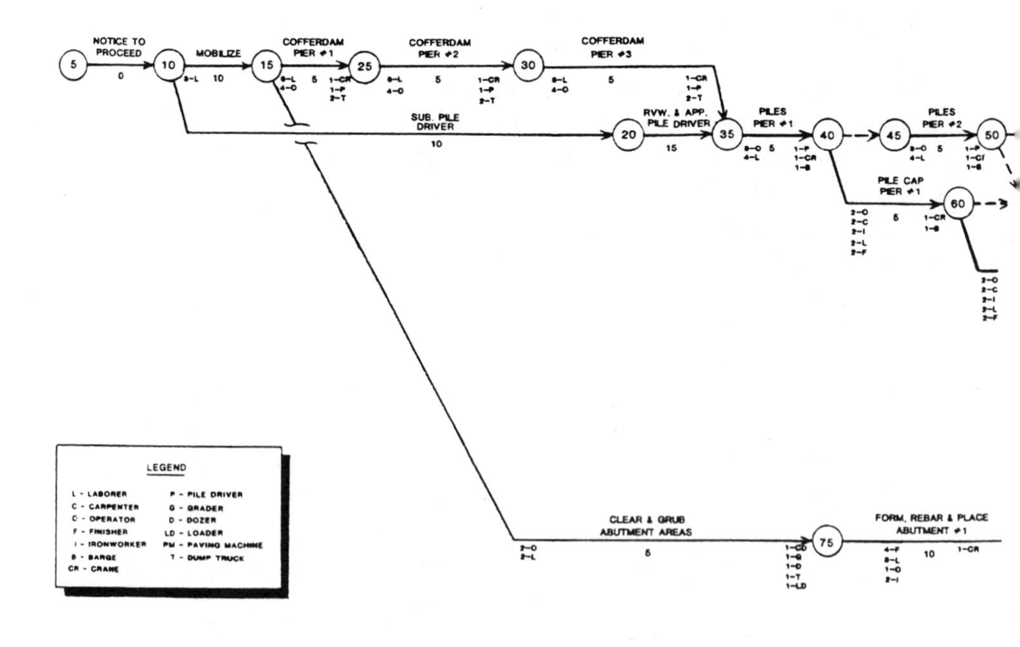

Figure 5-13. Logic diagram, with resources.

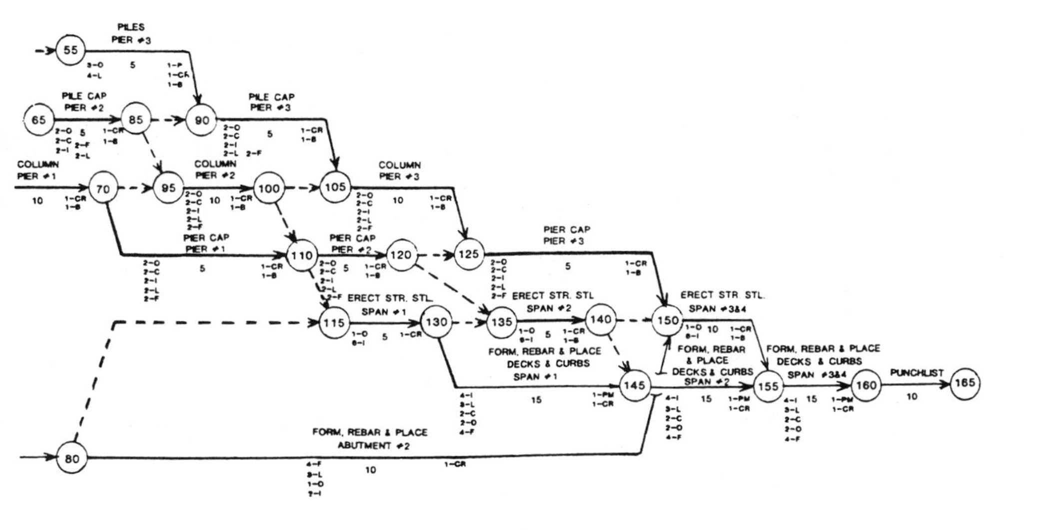

Figure 5-13. (*Continued*)

assigned for the amount of time allocated. For instance, if a contractor shows that it will install 30,000 pounds of reinforcing steel in a concrete deck in a period of three days with three ironworkers and one foreman, this might be considered a red flag regarding the reasonableness of the productivity for this work. In the same vein, the contractor can establish productivity goals for activities and monitor them to see if its planned efficiency is being achieved. The productivity is easily defined from the resource-loaded schedule. The same general advantages are gained for equipment. If the schedule shows that a crane is to be used for a specific activity in the next few days, and it is not yet on the site, this is an early warning to get the crane mobilized, rigged, load tested, and so on, to avoid delays. Finally, perhaps the most important area concerning the use of resources and float involves the concept of balancing or leveling resources.

Profits are maximized by utilizing your resources as efficiently as possible. Many factors can affect efficiency, such as the size of the crew, the consistency of the crew throughout the project, the amount of equipment needed for the project and maintained on the project site, and so on. By using the float in the schedule, we can optimize those resources. For the sake of discussion, we will demonstrate this with an example for labor in just one trade.

Figure 5-14 is an example of a logic diagram for a project with only carpenter labor assigned. Activity descriptions have been omitted to keep the example simple. No other trades and resources are shown—though they would be considered on an actual project. If we assume that all activities will start on their early start dates and add up the number of carpenters needed each day for the project, we would end up with the summary shown in Figures 5-15 and 5-16.

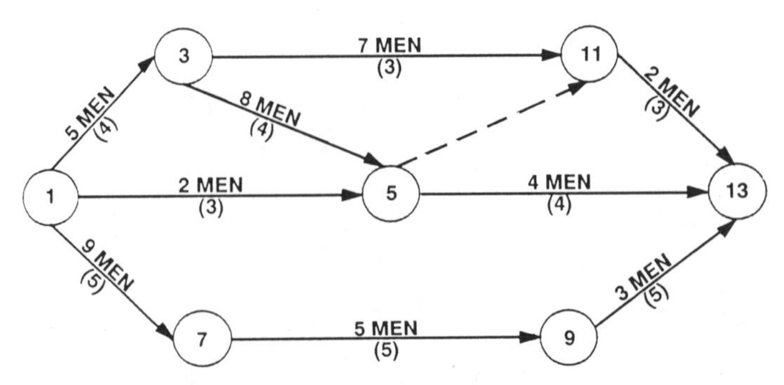

Figure 5-14. Logic diagram for carpenters only.

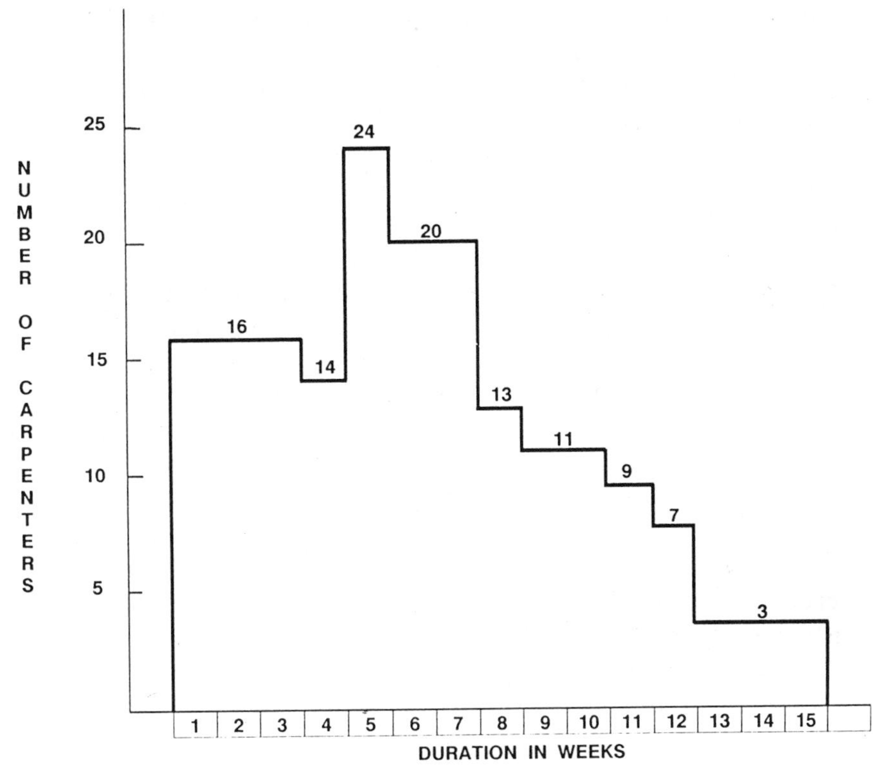

Figure 5-15. Carpenters graphed over project duration.

ACTIVITY NO.	1	2	3	4	5	6	7	8	9	10	11	12	13	14	15
1-3	5	5	5	5	X	X	X								
1-5	2	2	2						X	X	X				
1-7	9	9	9	9	9										
3-5					8	8	8	8	X	X					
3-11					7	7	7		X	X	X	X			
5-13									4	4	4	4			
7-9						5	5	5	5	5					
9-13											3	3	3	3	3
11-13									2	2	2				
TOTAL MEN	16	16	16	14	24	20	20	13	11	11	9	7	3	3	3

(NOTE: X represents interfering float.)

Figure 5-16. Distribution bar graph.

It is obvious from Figure 5-16 that our distribution of carpenters over the job is erratic. As managers, we recognize that we will achieve maximum productivity over the project by employing as small a crew and maintaining as constant a size crew as possible. We prefer to avoid variations in the crew size because it reduces the efficiency of the overall carpenter work. We may also have restrictions in terms of the size of the crew we will use. For example, perhaps as a general contractor we employ 16 carpenters full time throughout the year. If so, we should like to keep all of them busy and not have to hire additional help for short periods. Alternatively, as a larger union contractor, the local collective bargaining agreements may require certain nonworking supervisory personnel for different levels of craft personnel. In this case we would desire to keep our crews below these union threshold levels to avoid the extra expense of the supervisory personnel. For the sake of our example, let's presume that our goal is to have no more than 16 carpenters on the job at one time.

By using the float and moving the start dates for activities, we can level out our carpenter resource such that our goal is met of not exceeding 16 carpenters. This leveling is shown in Figure 5-17.

This same process can be performed for all trades and equipment. For instance, we might have an initial schedule that requires two cranes on the project for the same period of time. By shifting activities within the bounds of their float, we may be able to reduce our need to only one crane. This saves the cost of the additional mobilization and demobilization of the second crane. It should be clear now that the proper use of float can result in significant savings and increased profits for the manager.

ACTIVITY NO.	1	2	3	4	5	6	7	8	9	10	11	12	13	14	15
1-3	5	5	5	5	X	X	X								
1-5	2	2	2						X	X	X				
1-7	9	9	9	9	9										
3-5					-	8	8	8	8	X	X				
3-11					-	-	-	-	-	7	7	7			
5-13							-		-	4	4	4	4		
7-9						5	5	5	5	5					
9-13											3	3	3	3	3
11-13									-	-	-	-	2	2	2
TOTAL MEN	16	16	16	14	9	13	13	13	13	16	14	14	9	5	5

Figure 5-17. Leveling the resource distribution.

Finally, note that if we progress through a project allowing activities to slip past their early start dates, once we exhaust the available float, the activity becomes critical. This means that we have even more critical activities to be concerned with and an even greater risk that an activity will be late and adversely affect the scheduled completion date of the project. This is one of the reasons that free float was mentioned. If an activity has free float and we delay its early start, no other activities will be affected. If, however, the float is not free, when we utilize it on the first activity, we also reduce the available float for all succeeding activities in that path. This can lead to inadvertent problems for our subcontractors or our other resources. For example, if we allow one subcontractor to start late on an activity but within the amount of allowable float, this may reduce the float on succeeding activities involving other subcontractors. Subcontractor number two, which has had its float consumed, may now be unable to level its resources and, as a result, have higher costs. It is possible that because of limitations of resources, the project could be delayed.

Cost-Loaded Schedules. If we have developed a CPM schedule which defines time for activities and we have considered all resources for activities, we can review each activity to determine the materials involved. Once this is done, we can assign a cost value for each activity in the schedule. This results in what is known as a *cost-loaded CPM schedule*. This can be helpful to the manager, and the owner appreciates a cost-loaded CPM because it can be used to project the cash flow for pay requests during the course of the project. In fact, the owner could compare the payments on the project based on the assumed early start of all activities and the assumed late start of all activities. This would define a band of potential cash flow for the draws, as shown in Figure 5-18. Similarly, the contractor can determine its cash flow for the project.

The use of a cost-loaded schedule also reduces the amount of subjectivity involved when a pay request is submitted. Since we are dealing with discreet defined activities, it is much easier to determine if they have been completed or their respective percentage of completion. This significantly reduces the chance of front-end loading or incorrect billing and payment.

A cost-loaded schedule is also helpful in resolving problems if the project falls behind schedule or if it must be accelerated for any reason. When a project either falls behind the scheduled completion date or a decision is made to accelerate the work to finish early, this usually results in an exercise of overtime, increased crews, increased costs, and

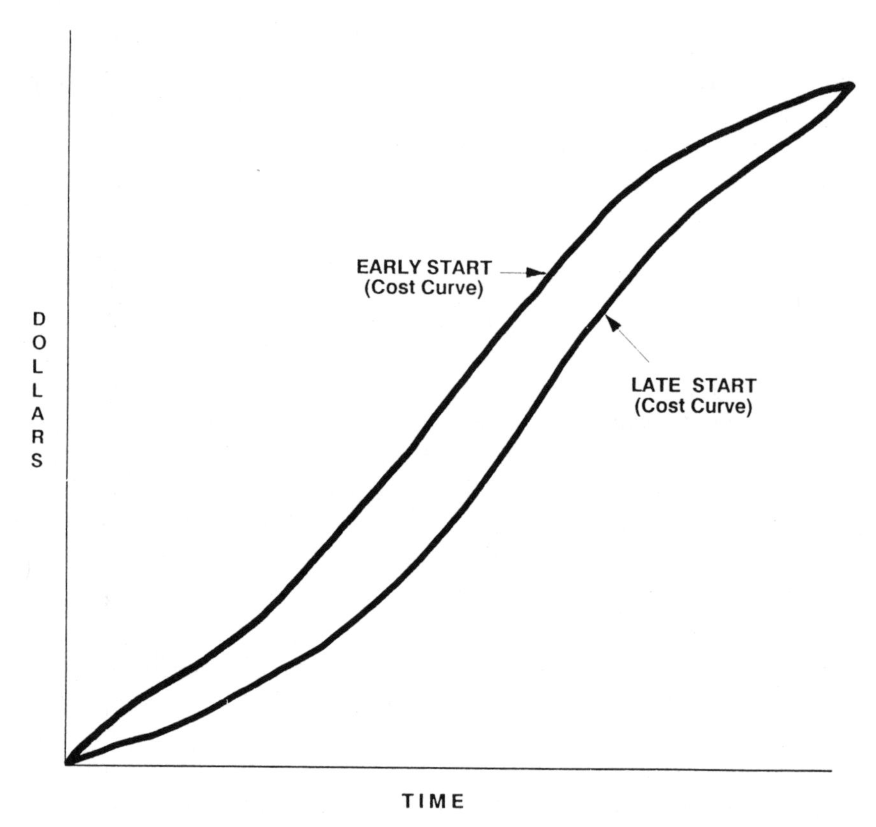

Figure 5-18. Spectrum of potential cash flow.

possibly no gain at all. The decision to accelerate or gain time should be made based on logical considerations of the plan and the progress. For example, assume that a contractor has fallen behind schedule because of its own problems. It must decide whether to accelerate in order to avoid the assessment of liquidated damages. How does it make the decision and which activities should it attempt to accelerate? In reality, the process is not as difficult or subjective as it may first appear.

Faced with a decision to gain time, the contractor would review the schedule to determine which activities remain to be performed and which are critical. Usually at this stage most remaining activities are critical. Having a cost-loaded CPM schedule, the contractor analyzes the remaining activities to determine the absolute minimum amount of time in which each can be performed presuming that it adds the necessary resources. This absolute fastest time is called the *crash time*. As it determines the crash time, it also determines how much each activity will cost in that crash time. This cost is called the *crash cost*. Then the contractor calculates the *cost slope* for the remaining activities:

$$\text{cost slope} = \frac{\text{crash cost} - \text{normal cost}}{\text{normal time} - \text{crash time}}$$

All of the elements of thise formula come from the cost-loaded CPM and the calculations for crashing activities. For instance, the contractor may have an activity that has been scheduled for 10 days at a cost of $5,000 and a second activity scheduled for 12 days at a cost of $4,000. In assessing the crash time and cost, the first activity can be crashed down to eight days at a total cost of $7,000 and the second activity crashed to eight days at a total cost of $10,000. The respective cost slopes would be:

$$\text{Activity 1: cost slope} = \frac{\$7{,}000 - \$5{,}000}{10 \text{ days} - 8 \text{ days}} = \$1{,}000/\text{day}$$

$$\text{Activity 2: cost slope} = \frac{\$10{,}000 - \$4{,}000}{12 \text{ days} - 8\text{days}} = \$1{,}500/\text{day}$$

Clearly it is less expensive to crash activity 1 than 2. For each day the contractor gains on activity 1, it spends an additional $1,000, while for each day gained on activity 2, it spends an additional $1,500. The contractor may determine after analyzing the cost slope of remaining activities that rather than accelerate, it is cost effective instead to accept the assessment of liquidated damages.

The application of this is similarly beneficial to the owner in deciding whether or not to accelerate a project. The process allows the manager to make a business decision as to the best course of action. No longer is it done solely by the seat of the pants.

MONITORING THE PROJECT

Though a manager may develop a terrific schedule for the project, unless it is used throughout the job, the effort was wasted. While it may have helped in the initial planning for the job, its most beneficial use is during the actual execution phase of the project.

To use the schedule to monitor the project, it must be updated at least on a monthly basis. Depending on the duration and nature of the project, more frequent updates may be necessary. An update of the schedule is a recording of the work that has already occurred on the schedule's activities. This allows the manager to determine if the project is ahead of, on schedule, or behind schedule. A typical update of our sample bridge project is shown in Figure 5-19. The update shows that the project is five days behind schedule. This is determined by looking at the final activity, Project Complete, and comparing the scheduled end date of October 13, 1993, with the presently forecasted end date of October 8, 1993.

In order to update the schedule, you must maintain a record of the work that has been performed, when it started, and when it completed. The duty for updating the schedule should be formally assigned to a staff member so that it is not forgotten. In the same vein, procedures should be established so that monitoring is an easy and routine task. One example would be to tie the project daily reports to specific activities on the schedule. Chapter 8 discusses this in detail. If changes occur during the update period, they must be incorporated into the updated schedule. This can either be in the form of revised logic, new activities, or changes to durations of existing activities. Also, the project manager should have periodic meetings to discuss the schedule and the progress, assess any problems, devise solutions, and if nothing else, keep everyone apprised of the status of the job with respect to time.

CPM scheduling is most easily performed with a computer and the associated software. But, while the computer makes the work faster and easier, everything the computer does, the manager could do by hand.

PRACTICAL CONSIDERATIONS FOR EFFECTIVE SCHEDULING

At this stage you may be at least a bit swayed in favor of using a CPM schedule for project planning and control. The enthusiasm can be quickly dampened, though, if executing the concept turns into a laborious task. The following guidelines will help you implement an effective scheduling program without creating additional problems and trauma.

1. *First time around*—If you are implementing the organization's first CPM schedule, it is suggested that it not be a complete resource-loaded and cost-loaded one. This could be a bit too complex for the first time around. To keep it simple, do the first CPM

just for time. Once you are comfortable with the technique, try a CPM with time and labor resources only. After this hurdle, make the next schedule with labor and equipment. The final step in the progression is the resource- and cost-loaded schedule.

2. *Paper for the portajohn*—Many managers enthused by the use of a CPM, produce a detailed schedule and provide Superintendents, foremen, subcontractors, and suppliers with reams of computer

TRAUNER CONSULTING SERVICES, INC. PRIMAVERA PROJECT PLANNER

REPORT DATE 3APR92 RUN NO. 24 START DATE 3MAY93 FIN DATE 13OCT93
14:10
SCHEDULE REPORT - SORT BY TF, ES DATA DATE 3JUN93 PAGE NO. 1

PRED	SUCC	ORIG DUR	REM DUR	%	CODE	ACTIVITY DESCRIPTION	EARLY START	EARLY FINISH	LATE START	LATE FINISH	TOTAL FLOAT
5	10	0	0	100		NOTICE TO PROCEED					
10	15	10	0	100		MOBILIZE					
10	20	10	0	100		SUB. PILE DRIVER					
15	25	5	0	100		COFFERDAM PIER #1					
15	75	5	0	100		CLEAR & GRUB ABUTMENT AREAS					
20	35	15	0	100		RVW & APP. PILE DRIVER					
25	30	5	0	100		COFFERDAM PIER #2					
30	35	5	5	0		COFFERDAM PIER #3	3JUN93	9JUN93	3JUN93	9JUN93	0
35	40	5	5	0		PILES PIER #1	10JUN93	16JUN93	10JUN93	16JUN93	0
40	60	5	5	0		PILE CAP PIER #1	17JUN93	23JUN93	17JUN93	23JUN93	0
60	70	10	10	0		COLUMN PIER #1	24JUN93	7JUL93	24JUN93	7JUL93	0
70	95	0	0	0		DUMMY	8JUL93	7JUL93	8JUL93	7JUL93	0
95	100	10	10	0		COLUMN PIER #2	8JUL93	21JUL93	8JUL93	21JUL93	0
100	110	0	0	0		DUMMY	22JUL93	21JUL93	22JUL93	21JUL93	0
110	115	0	0	0		DUMMY	22JUL93	21JUL93	22JUL93	21JUL93	0
115	130	5	5	0		ERECT STRUCTURE STEEL SPAN #1	22JUL93	28JUL93	22JUL93	28JUL93	0
130	145	15	15	0		FORM, REBAR, AND POUR DECKS & CURBS SPAN #1	29JUL93	18AUG93	29JUL93	18AUG93	0
145	155	15	15	0		FORM, REBAR, AND POUR DECKS & CURBS SPAN #2	19AUG93	8SEP93	19AUG93	8SEP93	0
155	160	15	15	0		FORM, REBAR, AND POUR DECKS & CURBS SPAN #3 & #4	9SEP93	29SEP93	9SEP93	29SEP93	0
160	165	10	10	0		PUNCH LIST	30SEP93	13OCT93	30SEP93	13OCT93	0
70	110	5	5	0		PIER CAP PIER #1	8JUL93	14JUL93	15JUL93	21JUL93	5
100	105	0	0	0		DUMMY	22JUL93	21JUL93	5AUG93	4AUG93	10
105	125	10	10	0		COLUMN PIER #3	22JUL93	4AUG93	5AUG93	18AUG93	10
110	120	5	5	0		PIER CAP PIER #2	22JUL93	28JUL93	5AUG93	11AUG93	10
120	135	0	0	0		DUMMY	29JUL93	28JUL93	12AUG93	11AUG93	10
130	135	0	0	0		DUMMY	29JUL93	28JUL93	12AUG93	11AUG93	10
135	140	5	5	0		ERECT STRUCTURE STEEL SPAN #2	29JUL93	4AUG93	12AUG93	18AUG93	10
125	150	5	5	0		PIER CAP PIER #3	5AUG93	11AUG93	19AUG93	25AUG93	10
140	145	0	0	0		DUMMY	5AUG93	4AUG93	19AUG93	18AUG93	10
150	155	10	10	0		ERECT STRUCTURE STEEL SPAN #3 & #4	12AUG93	25AUG93	26AUG93	8SEP93	10
120	125	0	0	0		DUMMY	29JUL93	28JUL93	19AUG93	18AUG93	15
140	150	0	0	0		DUMMY	5AUG93	4AUG93	26AUG93	25AUG93	15
40	45	0	0	0		DUMMY	17JUN93	16JUN93	15JUL93	14JUL93	20
45	50	5	5	0		PILES PIER #2	17JUN93	23JUN93	15JUL93	21JUL93	20
50	55	0	0	0		DUMMY	24JUN93	23JUN93	22JUL93	21JUL93	20
50	65	0	0	0		DUMMY	24JUN93	23JUN93	22JUL93	21JUL93	20
55	90	5	5	0		PILES PIER #3	24JUN93	30JUN93	22JUL93	28JUL93	20
60	65	0	0	0		DUMMY	24JUN93	23JUN93	22JUL93	21JUL93	20
65	85	5	5	0		PILE CAP PIER #2	24JUN93	30JUN93	22JUL93	28JUL93	20
85	90	0	0	0		DUMMY	1JUL93	30JUN93	29JUL93	28JUL93	20
90	105	5	5	0		PILE CAP PIER #3	1JUL93	7JUL93	29JUL93	4AUG93	20
75	80	10	10	0		FORM, REBAR, AND POUR ABUTMENT #1	3JUN93	16JUN93	8JUL93	21JUL93	25
80	115	0	0	0		DUMMY	17JUN93	16JUN93	22JUL93	21JUL93	25
80	150	10	10	0		FORM, REBAR, AND POUR ABUTMENT #2	17JUN93	30JUN93	12AUG93	25AUG93	40

Figure 5-19. Schedule update.

printouts and complex network diagrams. If these folks do not understand CPM—and many will not—the paper will only have use for the portajohn at the project site. Recognize the capabilities of the users of the schedule. If they do not thoroughly understand CPM, don't give them computer confusion. Use the CPM software to print out a simple bar chart of the activities that must be accomplished during the next two weeks or month. Print the bar chart based on early starts and explain to the users that they must work to accomplish all of the activities within the time frames set down in the bar chart. They should start on the day shown and finish by the end date. Only the manager has the need to know how to control the float. It is up to the manager to budget and allocate it as necessary.

3. *Updates and records*—In order to have the users develop the habit of tracking progress on the project, the manager can have them record the activities that are worked on right on the bar chart that has been issued. Figure 5-20 is a sample bar chart for two weeks of work on our sample bridge project. Figure 5-21 is that same bar chart with notes by the superintendent of the work that has been performed during the first week of that time period. With these, the manager can easily update the schedule at any point in time. It also keeps the staff keenly aware of the schedule, the activities to be done, and their actual progress on a daily basis.
4. *No computer, no problem*—Though most of us have been indoctrinated into the computer era, it may be best to do your first CPM schedule by hand. Though this is time consuming, it insures that you really understand the process. Anyone who has worked with CPM schedules on construction projects for a significant time knows that there are many individuals qualified to perform the mechanics of scheduling. But there are few people who really understand how to schedule effectively as a management tool.
5. *Size of project*—It is commonly asked what size project warrants the use of a CPM schedule. The answer is simple. A CPM schedule does not have to be used if a manager can keep in mind all of the project activities and their interrelationships. Few projects are this simple. And if a project is truly simple, so too is making and using a CPM schedule. Therefore, why not do it?

```
------------------------------------------------------------------------------------------------------------
TRAUNER CONSULTING SERVICES, INC.            PRIMAVERA PROJECT PLANNER

                                                                     START DATE  3MAY93  FIN DATE  8OCT93
REPORT DATE  3APR92   RUN NO.   21
        10:49
DAILY BAR CHART BY ES                                                DATA DATE  3MAY93   PAGE NO.    1

                                                                                    DAILY-TIME PER.   1
------------------------------------------------------------------------------------------------------------
............ACTIVITY DESCRIPTION.............        03    10    17    24    31    07    14    21    28    05    12
PRED   SUCC  OD   RD  PCT    CODES     FLOAT  SCHEDULE  MAY   MAY   MAY   MAY   MAY   JUN   JUN   JUN   JUN   JUL   JUL
---------- ---- ---- --- ------------ -----  --------  93    93    93    93    93    93    93    93    93    93    93
                                                    ----------------------------------------------------------------
NOTICE TO PROCEED                              EARLY    E     .     .     .     .     .     .     .     .     .     .
         0                          0                   *     .     .     .     .     .     .     .     .     .     .

MOBILIZE                                       EARLY    EEEEE..EEEEE .     .     .     .     .     .     .     .     .
        10                          0                   *     .     .     .     .     .     .     .     .     .     .

SUB. PILE DRIVER                               EARLY    EEEEE..EEEEE .     .     .     .     .     .     .     .     .
        10                          0                   *     .     .     .     .     .     .     .     .     .     .

COFFERDAM PIER #1                              EARLY    *     .     EEEEE .     .     .     .     .     .     .     .
         5                          0                   *     .     .     .     .     .     .     .     .     .     .

CLEAR & GRUB ABUTMENT AREAS                    EARLY    *     .     EEEEE .     .     .     .     .     .     .     .
         5                         30                   *     .     .     .     .     .     .     .     .     .     .

RVW & APP. PILE DRIVER                         EARLY    *     .     EEEEE..EEEEE..EEEEE .     .     .     .     .     .
        15                          0                   *     .     .     .     .     .     .     .     .     .     .

COFFERDAM PIER #2                              EARLY    *     .     .     EEEEE .     .     .     .     .     .     .
         5                          0                   *     .     .     .     .     .     .     .     .     .     .

FORM,REBAR, AND POUR ABUTMENT #1               EARLY    *     .     .     EEEEE..EEEEE .     .     .     .     .     .
        10                         30                   *     .     .     .     .     .     .     .     .     .     .

COFFERDAM PIER #3                              EARLY    *     .     .     .     EEEEE .     .     .     .     .     .
         5                          0                   *     .     .     .     .     .     .     .     .     .     .
```

Figure 5-20. Schedule broken down to bar chart.

SUMMARY

The single topic of scheduling includes many significant parts. Obviously, the manager has a choice of schedules to use. But to get the most benefit, the CPM schedule is unequaled. The use of the CPM, however, must be approached carefully. To use it effectively, the manager must understand it well, implement it correctly, and simplify it for use by the other members of the project team.

The project schedule is not solely a "time clock." In reality, it is an all encompassing plan. This plan can be presented in graphic (network) form and include manpower, equipment, and cost. Most importantly,

```
TRAUNER CONSULTING SERVICES, INC.              PRIMAVERA PROJECT PLANNER

REPORT DATE  3APR92  RUN NO.   20                                    START DATE  3MAY93  FIN DATE  8OCT93
            10:48
DAILY BAR CHART BY ES                                                DATA DATE  3MAY93   PAGE NO.   1

                                                                                  DAILY-TIME PER.   1
-------------------------------------------------------------------------------------------------------
............ACTIVITY DESCRIPTION.............            03   10   17   24   31   07   14   21   28   05   12
PRED  SUCC  OD  RD  PCT   CODES   FLOAT   SCHEDULE  MAY  MAY  MAY  MAY  MAY  JUN  JUN  JUN  JUN  JUL  JUL
---------- ---- ---- --- ----------- -----  --------  93   93   93   93   93   93   93   93   93   93   93
-------------------------------------------------------------------------------------------------------
NOTICE TO PROCEED                          EARLY     .    .    .    .    .    .    .    .    .    .    .
            0                    0                   *    .    .    .    .    .    .    .    .    .    .

MOBILIZE                                   EARLY     EEEEE..EEEEE .    .    .    .    .    .    .    .
           10                    0                   *    .    .    .    .    .    .    .    .    .    .

SUB. PILE DRIVER                           EARLY     EEEEE..EEEEE .    .    .    .    .    .    .    .
           10                    0                   *    .    .    .    .    .    .    .    .    .    .

COFFERDAM PIER #1                          EARLY     *    .    EEEEE .  .    .    .    .    .    .    .
            5                    0                   *    .    .    .    .    .    .    .    .    .    .

CLEAR & GRUB ABUTMENT AREAS                EARLY     *    .    EEEEE .  .    .    .    .    .    .    .
            5                   30                   *    .    .   NO WORK BY SUB .    .    .    .    .

RVW & APP. PILE DRIVER                     EARLY     *    .    EEEEE..EEEEE..EEEEE .    .    .    .    .
           15                    0                   *    .    .    .    .    .    .    .    .    .    .

COFFERDAM PIER #2                          EARLY     *    .    .    EEEEE .   .    .    .    .    .    .
            5                    0                   *    .    .    .    .    .    .    .    .    .    .

FORM,REBAR, AND POUR ABUTMENT #1           EARLY     *    .    .    EEEEE..EEEEE .    .    .    .    .
           10                   30                   *    .    .    .    .    .    .    .    .    .    .

COFFERDAM PIER #3                          EARLY     *    .    .    .    EEEEE .   .    .    .    .    .
            5                    0                   *    .    .    .    .    .    .    .    .    .    .
```

Figure 5-21. Bar chart containing manager's notes.

the effective use of the schedule as a management tool can help the project manager monitor time, track productivity, balance resources to maximize profits, and anticipate and react to changes, deviations, and problems. This is risk management at its best.

The proper execution of project scheduling and control does not come easy. It requires time and effort, but the rewards far outshine the investments.

CHAPTER 6

CHANGE ORDERS AND EXTRA WORK

One of the most challenging areas of management on the construction project is that of changes and extra work. Generally, when a project is in trouble, the problems stem from changes that have occurred. In order to effectively manage changes, the manager must thoroughly understand what they are, why they occur, and how they should be resolved.

Understanding alone, however, is not enough. The manager must be committed to resolve changes as quickly as possible. That commitment must be made on the part of the three parties involved in the contract: the owner, the designer, and the contractor. Resistance on the part of any one of the three can cause the resolution process to stagnate, thus increasing the chances of major problems.

Not fully understanding the change order process or failing to properly administer it will increase the risk that the change order process may overshadow the basic contract work. Risk also escalates for the chance of excessive change order costs, project delays, and, worst of all, major construction disputes.

The following discussion first addresses the causes of changes. Next it covers the area known as "constructive changes" since these seem to be all too common on projects today. The discussion then moves to how we must manage changes and the ramifications of time-related changes and associated costs.

CAUSES OF CHANGES

Simply, a change is any work required of a contractor or subcontractor that was not specified in the original contract documents. In other words, the contractor is being required to perform something different from what it agreed to.

There are many reasons why changes occur on a project. In the most general sense, no set of plans and specifications is perfect. There will always be some errors or omissions leading to changes. We must also recognize that construction projects today are more complex and sophisticated than in years past. This increases the opportunities for errors and misunderstandings.

The most common type of changes that occur include errors and omissions, differing site conditions, and postbid decisions on the part of the owner or user of the project concerning the physical characteristics of the job. Errors and omissions are normal and it is for this reason that a contingency should be allowed on all projects. As discussed earlier, the incidence of errors and omissions can be reduced by design and constructability reviews.

Owners differ in their philosophies concerning errors and omissions. Some owners accept them and pay for any work occasioned; others demand that the design professional pay for all associated costs. A middle ground is the best, most reasonable position. For example, suppose it is discovered during the course of construction that an item was omitted from the plans and specifications. The contractor will request additional payment for the cost to incorporate the item. If the owner were to demand that the designer pay for this, the owner would profit unjustly. For, if the item had been included in the original plans, the contractor's bid would have been correspondingly higher. This does not mean, though, that the designer has no liability. If the cost from the contractor to install the item is now higher than it would have been at the time of bid, then the designer might be responsible for the incremental cost. At the very least, the designer should perform any additional design work for no cost.

An owner who insists that the designer be responsible for all problems will then have a difficult time getting unbiased advice concerning whether or not an item is really a change. The natural motivation of the designer is to disclaim any change requests and attempt to put the burden on the contractor.

Differing site conditions have been discussed previously so they will not be rehashed in this chapter. Suffice it to say, that the more opportunities there are for site conditions to affect the work, the better the chance that differing site conditions will occur and lead to change orders.

Owner decisions or actions are simply the routine process of refining the project once it begins to take shape. The owner will realize during the course of construction that some items should be added, moved, or removed. Budget considerations may dictate changes during the course of the work. The owner may desire to occupy a part of the facility early and as a consequence, change the contractor's sequence of work. The list could go on at length. When an owner initiates an action that requires performance different from what was specified, a change occurs.

CONSTRUCTIVE CHANGES

A *constructive change* is any action or lack of action that alters the requirements of the contract. Constructive changes have caused a tremendous increase in problems on projects.

While a veritable "laundry list" could be drawn up of the various forms of constructive changes, examples help make concrete the *concept* of a constructive change. Managers who understand the concept will be able to assess if a constructive change occurs. Most importantly, if the concept is understood, the incidence of constructive changes occurring can be reduced.

A good example to start with is the constructive change known as *late inspection*. Presume that we have a project in which the specifications require the installation of underground pipe. The specifications further require that before the contractor backfills the pipe, it must be inspected by the owner's representative, and that the inspection will be made within 24 hours notice given by the contractor.

The contractor begins the pipe work and by Tuesday morning it is ready to backfill the first 200 feet of pipe. The contractor calls the owner's representative and requests an inspection. The inspection is not performed that day, nor the next, but is accomplished on Thursday, late in the afternoon. Is this a change?

Harkening back to the definition of a change—requiring something different from what is specified—this late inspection meets the criteria and is a change. The contract stated that the inspection would be performed within 24 hours of notice. The contractor gave notice and no inspection was performed within the period specified.

Merely because a change has occurred, however, does not mean that any additional cost is due. The second question that must be addressed is who caused the change. In this case, the owner did through the inaction of its representative. Had the contractor caused the change, no additional money would be due.

Once it is established that a change occurred and that the owner is responsible, the third and most difficult question to answer is what are the impacts of the change. An impact may be additional work, a delay, inefficient work performance, and others. If there is no impact from the change, then no money is due. For our example, presume that the contractor had a crew and equipment at the pipe trench site. After a few hours, the crew was moved to a new location to do other work. The equipment remained at the site. Given these facts, the contractor may have experienced the impacts of idle crew time plus time to demobilize and remobilize the crew plus idle equipment time. Given Murphy's Law, it might rain on Wednesday night, wash mud into the trench, and contaminate the bedding. As a consequence, the contractor would have to remove the pipe to replace the bedding and then reinstall the pipe. These impacts would merit additional compensation.

A few of the more common constructive changes include:

1. *Requiring a higher standard of performance*—Only the minimum standard of performance is required in accordance with the contract specifications. If an owner requires a contractor to perform above that, this may well be a constructive change. For example, a contract specifies that an area must be backfilled with on-site material and compacted with three passes of a sheepsfoot roller. The contractor backfills and compacts but the work is then not accepted because the soil does not pass a 90% modified Proctor test. This is an example of requiring a higher level of performance. According to the contract, the only standard that was required was three passes with the roller. To now require density to satisfy the modified Proctor test is a change. This type of constructive change is normally related to defective plans and specifications.
2. *Improper rejection*—This type of constructive change is closely associated with requiring a higher standard of performance. It normally relates to the rejection of some performance criteria of the contract with which the contractor has complied. For example, if the compaction noted in the above example is now rejected by the owner, it would be an improper rejection. Requiring the

contractor to compact to 90% density is the higher standard, and having it remove the lifts already completed with the three passes of the roller is an improper rejection.

3. *Failure to disclose material information*—Any information the owner has which may affect the contractor's method or manner of performance should be made available to the contractor. Failure to do so may be a constructive change. For example, a contractor performing a contract for dredging experiences materials that are extremely hard to dredge and almost rocklike. The material includes cobbles and boulders that were not apparent from the information in the contract. When the situation is addressed with the owner, it presents a set of soil borings that show exactly the conditions encountered. Unfortunately, these borings had not been made a part of the contract, nor were they available to the contractor at the time of bid. This was a constructive change of failure to disclose material information (sometimes referred to as "superior knowledge").
4. *Delays caused by action or lack of action on the part of the owner*—There are many ways that an owner can delay a project. For instance, if the contract specifies that certain shop drawings and submittals must be prepared by the contractor, and that they will be reviewed and returned within 30 days, failure to do the review in the time specified may result in a constructive change and a delay to the project. This can also apply to questions or clarifications requested by the contractor.

The list could go on. The most important element of a constructive change to recognize is that a requirement is made that differs from that which was specified and it is being effected not by issuing new drawings but rather by actions or lack of actions.

Why do constructive changes cause problems? The answer lies in the very nature of the cause. In our first example of late inspection, the cause of the problem was the tardiness of the owner's representative. To now accept that a change occurred, the owner's representative must admit that he or she made a mistake, that he or she was wrong. This is very difficult for anyone to do. As a consequence, there is a tendency to deny the situation as a change. This may well lead to a dispute. Similarly, for our higher standard example, the design firm must admit that it made a mistake in not specifying the 90% compaction requirement as opposed to the three passes of the roller. Again, there will be a tremendous reluctance to do that.

MANAGING CHANGES

How does a manager effectively prevent changes from becoming a project albatross?

The answer is simple. Read the contract, and comply with what is specified. This presumes, of course, that the contract was well written concerning changes.

Normally, the contractor must give written notice to the owner of any situation which it believes is a change. The notice must be given within the timeframe specified. Most importantly, the notice should be a clear, tactful letter that explains all the facts, the significant dates, the impacts to the project, and, if possible, the costs. This bears some additional discussion.

Many contractors believe that they should tell the owner as little as possible. They further contend that they cannot determine the costs and must wait until the work is actually performed. Both of these contentions are questionable. Problems cannot be solved until the owner has all the facts necessary. Therefore, it is in the contractor's best interest to lay out everything as soon as possible. Too, it is difficult to believe that a contractor cannot determine costs because the strongest ability a contractor has is the ability to estimate. This is what distinguishes a contractor from someone who knows how to build things, but cannot estimate. At the outset of the project, the contractor may have submitted a lump-sum bid based on only a few days of reviewing the plans and a very short site visit. Now when a change occurs and the contractor has the benefit of knowing the site, knowing the productivity of the specific workforce, etc., it suddenly loses the ability to estimate. Not true! A bit of thought and a lot of homework will enable the contractor to provide the owner with an estimate of costs such that a change order can be resolved.

Most construction contracts require that a written change order be executed by both parties before any additional work is performed. A well-written clause will have some mechanism to keep the work moving if a change order cannot be resolved expeditiously, be it a field directive, a time and materials order, etc. The judicious contractor should insure that it has complied with the contract and that the work is performed only after all contract requirements concerning changes have been fulfilled.

The owner and the designer have a significant obligation in resolving changes, also. They, too, must comply with the contract. They should expedite any actions which they are required to perform. If they believe that the requested change is not a change at all, then the contract mechanism for this situation should be employed. At the minimum, they should monitor the work performed so that they have a record of the labor, material, and equipment that was involved.

All parties to the project must be constantly alert for the occurrence of changes and must be prepared to resolve them as quickly as possible. The longer they sit with no resolution, the greater the chance that they will not be amicably resolved.

The manager must be particularly cognizant of changes that may be construed to be *cardinal changes*. Lately, this type of change has been discussed frequently. A cardinal change is a change that is outside the scope of the contract. The typical changes clause states that the "owner has the right to make changes within the scope of the contract." An owner does not have the right to make changes outside the scope as defined in the contract. If an owner makes a cardinal change, it can be considered a breach of contract. What are the ramifications of this?

First, let us look at some examples of what might be construed as a cardinal change. Presume that a contractor is constructing a highway project. The contract has defined unit prices for most items of work. The owner, during the course of the work, directs the contractor to perform some paving operations on another section of highway several miles from this project. This would likely be a cardinal change. The new work is not part of the defined scope of the contract. To require the contractor to perform work in accordance with the same unit prices is unfair, and the contractor would have the right to refuse to perform the work. Other types of cardinal changes can be much more subtle. For instance, a contractor was constructing a three-level prison facility for a state agency. The three levels contained maximum, medium, and minimum security requirements. During the course of construction, the owner initiated over 350 changes to the contract. The final project as constructed was an adult diagnostic treatment center for the criminally insane. Effectively, the combination of changes altered the scope of the contract.

When a cardinal change occurs, the contractor may either refuse to perform the work, or, if a dispute arises concerning costs, the contractor may recover its costs based on the principle of quantum meruit or the value of the work performed. Since the cardinal change is a breach of contract on the part of the owner, the provisions of the contract no longer apply.

All parties involved in the project should be careful not to attempt to use the "changes" clause of the contract to perform work outside the scope of the contract. If the work could most efficiently be performed by the existing contractor, then a supplemental agreement should be proposed.

THE IMPACT OF CHANGES

Every change order has two major components that must be defined in terms of the impact of the change: costs and time. Costs are virtually always addressed. Time, however, is not.

Change Orders and Time

If no time is involved for the change, then this should be stated in the body of the change order. Never leave the question of time for resolution at the end of the project. Assess and resolve it contemporaneously with the change order processing. Consider the example of the project on which the owner's representative procrastinated concerning the decision of time with respect to changes. During the course of the project, the owner and its designer initiated a significant number of change orders. In response to these, the contractor submitted proposals with a detailed breakdown for labor, materials, costs, and also a request for a specific number of days for a time extension. In virtually every case, the owner's representative approved the change order proposals for the cost portion but declined to accept the request for time. Instead it responded that it would wait until the end of the project to determine if any time was due the contractor and grant a time extension at that point. This proved to be a colossal error. The time that had been requested by the contractor in the numerous change order proposals totalled approximately 90 days. Had the change orders been written and executed in accordance with the proposals, the issue would have been resolved. The project actually finished 12 months late. At that point the contractor requested a time extension for the full 12 months and the costs associated with the delays, which were in excess of $1,500,000. A protracted dispute ensued, costing both sides considerable time and money. The problem easily could have been prevented.

The manager must assess the impact of time for every change, no matter how small. This assessment must be made both from the perspective of the current updated schedule and the standpoint of resources. For instance, the manager might review a change based on the schedule and note that it does not affect a critical activity. But a review of the resources involved shows that to perform the change, the contractor must utilize a specialized piece of equipment. This equipment is on the project site but is involved in performing critical work. If the contractor attempts to perform the change work, it will affect other critical activities by withdrawing the equipment to use on the change.

As a consequence, a delay might occur. After careful study, the contractor notes that the change can only be performed after the critical activities are completed and that, in this revised sequence, a time extension is warranted.

Consider thoroughly every possible impact of a change concerning the time on the project. This includes not only the impact of a small change, but also the cumulative impact of numerous changes. For example, a contractor constructing a complex process facility was issued several hundred change orders. Each change order by itself did not have a measurable impact on the time or schedule. Since most of the changes required pipefitter hours, however, there was a significant impact. Because of the location of the project and the unavailability of pipefitters, the cumulative hours required by the changes caused a significant delay. There was only so much overtime that could be worked. Given the limited resource of the pipefitters, a running total of the extra hours would have shown that the project duration would be extended. Unfortunately, the problem was not recognized until after the project was delayed.

Change Orders and Costs

The costs involved in a change order are invariably addressed by all parties. How they are handled is a function of what the contract requires concerning cost. Most changes clauses provide for alternate costing methods such as:

- Established unit prices
- A negotiated lump-sum amount
- Time and materials
- A decision by the Architect/Engineer

The better written the changes clause, the fewer problems will occur in deciding costs for a change.

In using the established unit prices, the manager must be cautious that they directly apply to the change that is being initiated. For example, if a unit price was established for excavation and the owner now directs some additional excavation, it would seem obvious that the unit price should be used. However, if the original unit price was based on mechanical excavation on a large scale and this change requires hand excavation, the unit price might not be appropriate. The better defined the original unit prices are, the fewer controversies will arise later on. Similarly, the use of if and where directed clauses discussed earlier will be helpful in this area.

Presuming no unit prices are applicable, then the contractor will be required to submit a cost proposal for the changed work. Although only an estimate, this proposal should include a detailed breakdown for labor, materials, equipment, overhead, and so on. The owner should review this promptly and work diligently to resolve the change costs before the work is done. But as we noted earlier, the contractor is most skilled at estimating, and this should not be a problem as long as the extent of the change is clearly and completely defined.

Some construction professionals maintain that using the actual cost of the work is better than the estimated cost. The author disagrees with this philosophy. If the contractor waits until the work is accomplished to submit the bill, the owner may well react: "I didn't know it would cost that much" or "If I had known that, I never would have directed the work" or "It couldn't have been that much—I'm not going to pay that." Now this is a major problem. The contractor has recorded the costs as actual costs and is not willing to negotiate the amount down. The owner refuses to pay what it believes is too much. The obvious outcome is some form of dispute and ensuing arbitration or litigation. The problem can be avoided by resolving the change before the work is performed.

Using time and materials is an acceptable approach when the situation does not allow for resolution of the cost of the change before the work is performed. This should only be used when the work must progress immediately. Too many contracts slip into a lazy form of change order management by just using time and materials for all changes. Both the owner and contractor must aggressively work to resolve the change orders before the work is done. If time and materials is used, then the guidelines of the contract must be followed concerning what costs are allowable and how the information must be tracked and recorded. If the owner had the forethought to have a well-written changes clause it should preclude any problems with time and materials types of changes.

DELETIONS

One other area of change order costs that usually creates concern is that of *deletions* of contract work. When an owner reduces or deletes work, it is often difficult to agree on the cost that should be credited. This happens because the owner expects the deletion will receive a credit that matches the amount shown on the project schedule of values. The contractor's submission, however, may be far less in recognition that the schedule of values might not be an accurate representation of the cost of the work for the specific items. An easy way to minimize this potential for dispute is to use the cost-loaded CPM schedule discussed earlier.

The basic principle that should be adhered to for deletions is that the contractor should offer and the owner should expect cost that represents the amount the contractor would have spent to actually perform the work. This can be supported by a detailed breakdown of labor, material, equipment, etc., just as is normal for an additive change on a lump-sum basis. But what about overhead and profit? It is not reasonable that the contractor should be entitled to profit for work that it is not going to perform. Therefore, the owner should expect a credit for profit to be included in the change order proposal. Overhead, on the other hand, is somewhat different. If the contractor has burdened the job with overhead that is time related, then no credit may be due for the deletion. Since the overhead remains the same, presuming that the deletion does not reduce the contract duration, the contractor will not have any savings to offer the owner. In fact, it could be argued that the very processing of the change increases the contractor's administrative efforts, and hence, increases its overhead cost. As a general guideline, and dependent on the specific requirements of the changes clause, the owner should expect a credit for profit but little to no credit for overhead.

COST OF CHANGES

In managing changes, it is important to the manager to pay careful attention to all changes that are executed on the job. For example, presume that the owner initiates a small change almost at the inception of the job. The contractor submits the cost breakdown shown in Table 6-1.

TABLE 6-1

Labor—Carpenters (2 Hrs)	$32
Material—Plywood	$ 8
Small Tools	$10
Overhead	$10
Profit	$10
Total:	**$70**

Given the size of the change, it very likely will be approved. This same type of small change occurs several more times during the initial stages of the job. Then a much larger change is initiated. The contractor's cost proposal is submitted and looks like Table 6-2.

At this stage, the owner and its representative sit up and take notice. Many objections are raised to the amount for small tools, overhead, and profit. And these objections are reasonable. But the basic question remains: what does the contract say? If the contract changes clause does not specify anything concerning small tools, overhead, or profit, then the parties may well have established the reasonableness of the percentages used by the earlier changes that were mutually executed separately. Even if the contract does address these percentages or the application of these items on a change, having executed several changes with a different approach may well be construed to be amending the contract by virtue of the actions of the parties.

This same scenario could occur in the opposite way. The contractor's first submittal might have been marked down significantly by the owner. Instead of the $70, it was executed as a change for $42. The marked-up version looked like Table 6-3. If the same several early changes were all done similarly, the contractor may find that, despite the contract, new guidelines have been established concerning the management of change orders.

TABLE 6-2

Labor—Carpenters	400 Hrs.	$ 6,400
Masons	800 Hrs.	$ 16,000
Laborers	900 Hrs.	$ 10,800
	Subtotal:	$ 33,200
Material—Block		$ 22,000
Lumber		$ 10,000
Mortar		$ 12,000
Rebar		$ 9,000
	Subtotal:	$ 53,000
Small Tools		$ 21,550
Overhead		$ 21,550
Profit		$ 21,550
Total:		**$150,850**

TABLE 6-3

Labor—Carpenters (2 Hrs.)	$32
Material—Plywood	$ 8
Small Tools	$ 0
Overhead	$ 1
Profit	$ 1
Total:	**$42**

This example is not meant to encourage any gamesmanship on the part of any party to the project but to highlight the importance of every change, however small.

The costs of the direct work involved in a change are relatively easy to ascertain or estimate. But changes may involve indirect costs, which are more difficult to pin down. These generally occur with delays because of the extra work or the constructive changes previously described. Before we address the cost elements of time, an initial discussion of delays will be instructive.

DELAYS

A well-written contract will address delays. The schedule should be the tool used to measure these delays. In general, there are two types of delays: excusable and nonexcusable. *Excusable delays* are unforeseeable and beyond the contractor's control. As a consequence, the contractor is entitled to a time extension. *Nonexcusable delays* are those which were foreseeable or within the contractor's control. Examples of nonexcusable delays include those by subcontractors or suppliers, slow performance by the contractor, lack of materials, and late submissions. This type of delay may cause the contractor to be liable to the owner for liquidated damages or actual damages.

Excusable delays are further divided into the categories of noncompensable delays and compensable delays. *Noncompensable delays* are those for which the contractor is entitled to an extension of time but no additional cost. *Compensable delays* are those for which the contractor is entitled to an extension of time and additional compensation. Delays that are noncompensable must be defined in the contract. Typically, these include, strikes, fires, floods, acts of god, and unusually severe weather. The reason that a contractor receives a time extension is that these were unforeseeable and beyond its control. No money is due, however, because these problems were likewise not within the control of the owner. Hence, it is a form of "no fault." The owner absorbs any actual or liquidated damages by granting the time extension, and the contractor absorbs its indirect time-related costs.

Compensable delays are those caused by the owner. They may be related to directed changes in the work or to constructive changes. For these, the contractor is entitled to both time and money. The compensation due a contractor because of delays can involve many elements. The more common ones include field overhead, escalation, idle or extended labor and equipment, and home office overhead.

Field overhead encompasses the time-related expenses of maintaining the project site. This would include such items as the trailers, the superintendent, the clerk, and the portajohn. These costs are readily ascertainable since the items are present on the site and can be verified easily by the parties.

Escalation may affect labor, material, and equipment. A contractor must show when the respective craft labor rates increased and how the owner-caused delay forced a specific amount of hours into the more expensive timeframe. The amount of hours moved into the escalated period times the amount of escalation per hour would be the measure of the costs due. This same approach applies to equipment and materials. The contractor would have to demonstrate that the equipment rental rate increased and show which pieces of equipment were pushed into the more expensive period. The hours times the increase yield the costs. For materials, the contractor might show that the cost per yard of concrete from the suppliers increased at a specific date. The delay by the owner forced the contractor to place the concrete later, in the more expensive timeframe. Hence, the escalation cost would be due.

Idle or extended labor and equipment is similar to escalation. The contractor would be required to show that because of the owner-caused delay, it had equipment sitting idle on the site or was forced to use it for an extended period of time. The cost attributable to this would be included in the change order proposal.

Home office overhead is the most difficult area in which to determine cost because of a delay. Home office overhead represents the costs of supporting a home office such as the rent, utilities, staff, accountants, and so on. Most contractors prorate these costs to their various projects. If one project is delayed, it tends to underabsorb its share of these fixed home office costs and thus either to force the other projects to overabsorb the fixed costs or to become a loss to the contractor. The area of unabsorbed home office costs could entail a chapterlong discussion. Interested readers are advised to consult other texts containing more detail on this topic. Suffice it to say that in the face of an owner-caused delay, the contractor may be entitled to additional compensation for home office overhead costs. There are several approaches to calculating this, but the method used should be based on further readings and a clear understanding of the specific contract in question.

ACCELERATION

Another time-related change, which is the reverse of delays, is *acceleration* of work on a project. In some instances, an owner may desire that the project be finished faster than originally specified in the contract. Consequently, the owner may request a cost proposal for accelerating the project. This is a change just like any other. It requires a detailed cost proposal from the contractor and review by the owner. The elements of the proposal may include costs for overtime, second shift differentials, additional supervision, additional equipment, changes in materials, etc. A far more subtle change associated with acceleration is known as *constructive acceleration*.

Constructive acceleration occurs if all six of the following conditions exist:

1. The contractor has experienced an excusable delay, either noncompensable or compensable.
2. The contractor has given the owner notice of the delay and requested an extension of time.
3. The owner has refused to grant the time extension.
4. The contractor is directed to complete the project on time or in accordance with the then appropriate contract completion date.
5. The contractor accelerates its work on the project.
6. The contractor incurs additional costs or damages.

Note that there is no requirement that the contractor finish on time. Merely because it has accelerated does not mean that it can make up the time and finish without the benefit of a time extension. In reflecting on this situation it is easier to conceptualize constructive acceleration by comparing it to directed acceleration. If the contract completion date was October 5, 1993, and the owner requested the contractor to accelerate the project to finish by September 1, 1993, this would be a change and a cost could be negotiated. Conversely, if the original contract completion date had been September 1, 1993, and the owner had caused a delay such that the new date was October 5, 1993, then the contractor would be looking for a time extension to that date. By not granting the extension and instead directing the contractor to finish by September 1, the exact same set of facts exist.

The astute owner avoids the chance of constructive acceleration occurring by closely monitoring the schedule, determining when delays occur, ascertaining the liability for the delay, and granting time extensions in a timely fashion should they be due. The contractor has the same responsibility but its action is to request the time extension in a timely fashion.

SUMMARY

This chapter has addressed changes from several perspectives. It should be clear from the discussion that changes are a major element in the successful management of a construction project. All parties involved in the project must be keenly aware of this and must dedicate the resources necessary to properly manage changes. This requires that the staff be properly educated, be alert to changes, and be tenacious in the pursuit of resolving them as quickly as possible. Any failures in this area can lead to significant problems and disputes in which all parties may lose.

By properly managing the change order process, we are exercising control of the risks associated with budget and time. Maladministration of changes can result in unnecessary additional costs and possible delays to the project. During the construction phase of the project, the manager's focus is not on reducing the risk of changes since that should have occurred much earlier. Instead, the manager is attempting to limit the risk of excess costs and delays stemming from the change order process.

CHAPTER 7

INFORMATION MANAGEMENT

Seasoned construction professionals often complain about paperwork. They will cite the good old days when all the forms, reports, and other required documentation were not used. Though this might have been the way it was, it no longer is the case. Therefore, instead of lamenting the times past, the astute manager prepares for the proper management of information from the beginning of the project and monitors it throughout the project.

Even in an area as mundane as documentation and recordkeeping, the astute manager exercises due diligence with respect to risk. Numerous examples support the risk management aspect of prudent documentation. For example, if a project designer orally directs a contractor to change the size of reinforcing steel to a number 10 bar but this is misunderstood to be a number 2 bar, the results could be catastrophic. Similarly, if the owner's representative makes agreements concerning extra work but leaves the project before a formal modification is issued, the contractor may not receive the compensation due.

The prudence and caution associated with professional information management is necessary and the reduced risks are well worth the investment of the time required and an absolute necessity on a modern construction project. There are many reasons for this. The following list identifies the more common reasons but this list by no means should be considered all inclusive.

1. Appropriate documentation allows future users of the project to verify how the project was constructed.
2. Lessons learned on this project are chronicled so that future projects can benefit.
3. If problems occur on the project in the future, the search for causes will be greatly aided by complete project documentation.
4. Continuous and contemporaneous documentation reduces the chance that misunderstandings will occur concerning the day-to-day project communications.
5. The sheer volume of information on a construction project exceeds that which any one individual can retain. Therefore, maintaining records precludes the loss of important information.
6. Should disputes arise during or after the construction, the contemporaneous documentation may be the best, and, in some cases, the only means to sort out the "truth."
7. If any turnover occurs in project personnel, the replacement staff will have a record of what has transpired.
8. Contemporaneous documentation is an efficient means of keeping multiple parties informed of the status of the project and reduces the number of meetings required and the amount of oral communications or reports.
9. Recordkeeping is invaluable in the monitoring process. By focusing your required documentation on areas that must be monitored, you can effect your measurement and updating process far more efficiently.
10. The basic step of establishing defined documentation assists the manager in focusing on the most important aspects of the project.

PLANNING YOUR RECORDKEEPING

The process of recording information cannot be impromptu. The manager who approaches recordkeeping with a "do it as appropriate" attitude will find either that too many bits of useless information will be recorded or that key data will not. At project inception, the manager should give careful thought to the types of documentation that will be important to the project, how this information will be recorded, who will maintain it, and how it will be utilized and disseminated.

In planning for an efficient information management system, the manager should begin by listing the types of information that will be needed just for routine operations. For instance, the first list might look something like this:

1. Daily personnel
2. Equipment on site
3. Daily work accomplished
4. Meeting minutes
5. As-built records
6. Correspondence
7. Requests for information
8. Status reports
9. Cost records
10. Specifications and deviations
11. Weather conditions
12. Problems that arise
13. Visitors to the site
14. Payrolls
15. Memoranda
16. Test reports
17. Inspections
18. Photographs or videos
19. Progress of the work
20. Field communications

This type of list is a starting point. Now the manager must determine the purpose of each type of record, establish the format for the record, and determine the individual best able to record the information. For instance, the manager needs a record of the daily work accomplished. This information helps in updating the project schedule and in determining an appropriate amount for periodic invoicing. If our manager is the owner's representative, he or she might then determine that the chief inspector should record this information and that the contractor will be required to submit the same information each day. Once that decision is made, the manager can then decide on an appropriate format for recording the information. Finally, he or she must decide who reviews the information, and how it will be filed and maintained, and who receives copies.

Since the process is repetitive for successive projects, the manager probably can utilize the systems established for prior jobs in the present one. As one gains experience, the procedures are modified and refined to suit the individual needs of the job.

The following discussion presents typical records that are needed on a project. Additionally, it explains why they should be maintained and provides some ideas on how to accomplish them. In that context, examples of both good and bad documentation will be illustrated.

TYPES OF AND REASONS FOR DOCUMENTATION

The requirement to maintain records should begin at the very inception of the project. If the owner provides guidelines during the early design stage, these should be documented both by the owner and the design professional. In many instances, this type of information is given orally during meetings among the two parties. Both should follow up the meeting with a letter or memorandum noting the discussion and any significant decisions reached. For instance, if the owner expresses a desire to have gold-plated fixtures, this should be documented. While the owner may believe that it clearly expressed this as a requirement, the designer may have interpreted this as a wish. Follow-up communication will clarify the point and eliminate the chance of a misunderstanding.

Design Documentation

Numerous meetings will be held as design documents are developed. These meetings should be documented. A good example of the importance of this was a project that involved the design and construction of a new highway. During an early design review meeting, the owner suggested that the designer include a construction joint in a portion of the elevated roadway. The owner further noted that the designer should verify the feasibility of this and make any necessary changes. Since Murphy's law always applies, the designer included the construction joint but did not pursue further studies concerning its use or ramifications. During construction, that specific section of the elevated roadway collapsed, causing both physical injuries and additional costs. This led to a protracted and expensive lawsuit. During the litigation it was noted that there were no records of the meeting and as a consequence each side relied on their own recollections of what was said and what was intended. A simple letter could have prevented this situation from occurring.

As a general guide, all documents developed during the prebid stage should be maintained at least until the completion of the project. This would include preliminary drawings, meeting minutes, addenda to the contract, estimates, quotations, and assumptions made. While we cannot discuss every type of record and the reason for it, let us use another example to illustrate the importance of the prebid documents.

Assume that during the course of a project the owner decides to change the specifications concerning specific equipment to be installed. The contractor is advised that the owner now desires a specific brand of pump. A change order is initiated and the contractor must supply a cost breakdown for the change order proposal. In its cost summary, the contractor requests additional money for the pump, asserting that it bid the job based on another brand of pump that was less expensive. To support that position, the contractor should be able to present a copy of the quotation it received at the time of bid and also any applicable bid documents that support its use of the brand that it had based its bid upon. All too often, this information is not available since it was either not created or it was not maintained. As a consequence, it may be difficult for the owner and the contractor to reach an agreement on the cost for the change. By establishing a procedure for written quotations and for noting specific brands in the bid documents, the problem could have been avoided.

Once a project is underway, it is common for revisions of the drawings to be issued. Many parties will then dispose of the older version of the drawings based on the belief that this will preclude any opportunity for confusion regarding the current drawing. Unfortunately, revised drawings usually contain only a brief note as to the nature of the revision. Should problems occur, it may be necessary to track back through the evolution of the drawing to determine exactly what was changed. Though we do want to avoid confusion, it is a simple procedure to stamp an outdated drawing as "not for construction use" to indicate that it is no longer applicable. The outdated drawing should then be filed.

It may be wise to maintain a drawing log which records by number all drawings and revisions. The log should also record when the drawing was issued and received. Periodically, the log can be reviewed by all parties to insure that all parties are up to date. Some managers establish a policy of date-stamping drawings when they issue or receive them.

Daily Reports and Records

One of the most important documents on a construction project is the daily report, which should include the location of the work, the work accomplished, the crew utilized, the hours worked, the equipment used, the weather, and any significant events or problems that occurred. Daily reports should be dated and signed by the individual recording the information. While this sounds obvious, many daily reports may lack either dates or signatures.

All this detailed information helps the manager in several ways. First, the manager needs to track progress of the work in accordance with the schedule. Without recording the work on a daily basis, one would have to periodically walk the site and estimate the amount of work performed. It is far easier and more accurate to record this information each day. If a CPM schedule is being used, the daily report should tie the work to specific activities on the schedule. This avoids any ambiguities as to the work that was performed, and it greatly facilitates the updating of the schedule.

An example of a project daily report is shown in Figure 7-1.

A quick review of the report in Figure 7-1 shows that the information is vague and may have little meaning to someone who is not fully involved at the site. A more detailed version of this report is shown in Figure 7-2. This amount of detail might seem too time consuming, but a moment's reflection should dispel this concern. In a normal project, only a limited number of activities are worked on any given day. Therefore, the effort to be specific will only take a few moments more than the vague references noted in Figure 7-1.

The report in Figure 7-2 is specific as to the work accomplished and notes the CPM activities by their respective node numbers. The more refined daily report also is specific in the narrative of the work accomplished. It refers to the exact location of the work as opposed to the very general category of "pile driving." The crew sizes are clearly noted, as well as equipment used and time expended. This information can be extremely useful if problems arise concerning the productivity of the work performed. Also, it provides the contractor with daily feedback on the rate of production such that it can be measured against the budgeted rate of efficiency. For example, a State Department of Transportation was sued by a contractor for a differing site condition concerning the pile driving on piers for a major highway bridge. The contractor alleged that because of differing soils conditions, the pile driving for seven of the piers was less efficient. The state inspectors had kept detailed records of the pile driving operations and were able to

Daily Report

Date 6-3-92 Contract C-69392

Work Hours 7.4 Weather Clear

Work Accomplished:

Pile driving, cofferdams, dewatering

Signature Joe Smith

Figure 7-1.

show that the pile driving productivity for the piers with the alleged differing site condition was equal to or better than the production for the remaining piers. The inspectors' records were detailed to the minute for each pile, recorded the exact pile length, the efficiency of the hammer, the length of any cutoffs and splices, and the time it took to cutoff or splice. The diligence on the part of those inspectors saved the state a significant amount of money.

Furthermore, a good daily report notes any significant events, records all equipment on the site, and chronicles the presence of visitors. Anyone who has been involved in a major dispute concerning delays or productivity will recognize the tremendous benefit that records of this type can provide.

In a project's planning, the manager will most likely develop a standard form such as that shown in Figure 7-2. Developing the form is not enough; the manager should insure that the staff that will be utilizing the form understands how it is to be filled out and what is expected.

Daily Report

Date JUNE 3 1992

Report No. 97

Contract No. C-69392

S M T W T F S

Job hours from____ To______

Clear	Partly Overcast	Overcast	Rain	Snow
to 32°	32°-50°	50°-70°	70°-85°	85° up
Wind	Calm	Light	Moder.	High
Humidity		Dry	Moder.	Humid

Contractor	Crew by Trade	Equipment	Working	Idle	Comments
Ace Const.	4 Carp.	1 Crane	8	–	
	6 Lab	1 Pile Driver	4	4	Had problems with Leads
	3 oper.	1 grader	2	6	
Smith Concrete	6 Carp.	1 Concrete Pump	6	2	
	6 Lab.				

Work Progress

Activity	Work Performed
15-20	Completed at 2:20 p.m. Grade beams JK2 + JK3
20-25	Started forming footings A1 + A2
10-35	Graded SW corner of site
5-45	Drove 11 piles for WT foundation

Signature Joe Smith

Figure 7-2.

Then the manager must review the reports as they are prepared to be sure that the requirements are being met. This concept will be discussed further under the topic of procedures manuals.

Some construction representatives use pocket tape recorders to supplement daily reports. At the end of the week, the tapes are transcribed and effectively become the daily report. Although this approach is not normally necessary, circumstances sometimes dictate its use. For example, a recent "crisis" project required 3 shifts, with work proceeding 24 hours a day. Because of the pace of the work, tape recorders were used.

Whether in the daily report or some other form, a record should be maintained of the project equipment and its use. This could be important to both the owner and the contractor. For example, a governmental agency received a claim from a contractor for delays on a project. Part of the damages claimed were costs for idle equipment. Since the agency had maintained an accurate record of the equipment usage, it was able to demonstrate that most of the equipment was idle both before and after the suspension of work that had occurred. As a consequence, the contractor was not entitled to additional compensation for that equipment. Contractors may often find that their site staff has a tendency to hoard equipment "in case they need it." By monitoring equipment usage, the contractor may be able to more efficiently allocate its resources among multiple jobs and educate the staff to request equipment only when it is really needed.

It is also advisable to maintain a record of visitors to the site. This can be done in the daily reports or a separate sign-in log. This record would be important should questions arise as to the frequency of the designer's visits or the presence of government inspectors, for instance.

Many site staff personnel maintain diaries in which they record significant events, discussions, conversations, and so on. The use of diaries is an individual decision for the manager. The same information can be maintained in many other forms such as memoranda or letters. It is commonly believed that diary should be kept in a bound book to avoid the suspicion that, if in loose leaf form, information could be added after the fact and inserted into the diary. In fact, the actual form of the diary does not matter. If diary information becomes important, particularly in a legal sense, no problems should exist as long as the information can be authenticated.

CORRESPONDENCE AND MEMORANDA

Letters should be written to clarify communications and to inform parties absent from the job site of significant events. Because oral communication is often misinterpreted, a simple guideline for the manager is to write a letter when giving or receiving an oral directive or reaching an oral agreement concerning the conduct of the work. This process does not have to be tremendously time consuming. It is entirely acceptable to use handwritten speed memos to clarify discussions. An example is shown in Figure 7-3.

Memos should be written to personnel on the manager's staff or to the file. Memos to the staff keeps it informed, provides direction, and clarifies actions, among many other reasons. But why write memoranda to the file? Memos to the file document important events or decisions. They insure that things are not forgotten and if new staff members become involved in the project, the file provides an adequate base of information from which to work. If you have taken over a project during the course of construction, you probably first reviewed the file to get "up to speed." You probably found, as is true in the vast majority of cases, that the file was inadequate.

Because of the litigious nature of construction contracts, a trend has developed to write letters and memos for self-protection. This practice should be discouraged. A document written for protection normally sounds just like that, a self-serving document laying blame at someone else's feet. Exhibit 7-1 is an example of self-serving writing.

carbonless DUP 9:4 redi-note®

REDI-NOTE DATE June 4 19 92

TO Bill Jones, Ace Const.

SUBJECT To clarify our discussion today, though the shop dwgs. showed otherwise, the reinforcing pattern for the footings, A1, A3 + B3 can be the #8 bars at 6". As noted, this is a no cost change.

SIGNED Joe Smith, RE

REDIFORM® 4S462
POLY PAK (50 SETS) 4P462
carbonless

☐ NO REPLY NECESSARY ☐ REPLY REQUESTED - USE REVERSE SIDE

Figure 7-3. Job site speed memo.

The same information could have been written in the context that was previously discussed. This is shown in Exhibit 7-2.

Memo

To: File

From: Joe Smith, Resident Engineer

Date: 11/12/92

RE: Contractor Performance

I noted that the Contractor's staff is totally incompetent when it comes to reading plans. During my site walk I was barraged with the most incredibly basic questions concerning the layout of the reinforcing steel for the shear walls. It was clear to me that the contractor either doesn't know how to read drawings or is starting to set up the owner for an extra.

I answered all the questions and though it is clear from the drawings, I'm not sure they understood or will do the work correctly.

Exhibit 7-1

Memo

To: File

From: Joe Smith, Resident Engineer

Date: 11/12/92

RE: Shear Wall Construction

The Superintendent for the contractor raised questions today concerning the fit of the reinforcing steel at the junction of the shear walls between the perimeter and the elevator shaft. The detail shown on drawing S-3 is confusing. I showed the Super that sheet S-7 details resolved the problem at detail C.

I believe the problem is resolved but will monitor the actual work to insure compliance.

Exhibit 7-2

Because we often communicate important information through telephone conversations the astute manager will maintain a record of telephone calls or a phone log. Once again, this insures that important information is not lost. An example of a telephone log is shown in Figure 7-4.

Telephone Log *Name* Joe Smith

Date	*Telecon With*	*Subject*
June 4, 1992	Bill Jones (Ace Constr.)	Bill called and agreed that bearing pads are at the wrong elevation. He said Ace will jack hammer the supports and correct this.

Figure 7-4. Example phone log.

PHOTOGRAPHS

Every construction project should have a camera at the site. Normally, periodic photographs are required on a project. They should also be taken to record such significant events as flooding of the site, accidents, or incorrect workmanship.

Though in most instances a 35mm camera is adequate for pictures, a Polaroid camera should also be available. If an event must be documented by a photograph, you would not want to be surprised when the film is returned blank. To insure a record, you can take a Polaroid photo and then take 35mm shots to be developed later.

When pictures are taken, they should be dated and signed. The dating can easily be accomplished with the use of a date stamp camera. You can't avoid the process of signing the back of the pictures once they are developed. The signature may be necessary should a problem arise. The signature on the back authenticates the photos.

It is also helpful if a brief description is written on the back of the photo. This helps to clarify exactly what the picture is attempting to show.

When photographs are taken, they should be meaningful. For instance, suppose a major storm floods the area on site that has been excavated. By taking a photo that shows only brown water, it is difficult to determine the extent of the problem. The photo needs a perspective.

This can be achieved by inserting a hard hat or a ruler to provide a scale, by including background landmarks or other points of reference.

Photographs are an excellent way to record progress on the project. This is the reason many contracts require that monthly photos be taken. To enhance the effectiveness of this method, it is useful to correlate the pictures taken with the schedule on the project. If the schedule reports specific activities complete, photos can be taken to support this.

Recently managers have begun to use videocassettes to record progress or to chronicle significant events. This is an excellent method of documenting information. Remember, however, that a videocassette records everything that is said. Therefore, the person filming the project should report information factually and avoid any innuendos or aspersions concerning the work or progress.

SCHEDULES AND UPDATES

All schedules prepared for the project should be maintained, as well as all updates of the schedule. Should any problems occur concerning progress, the schedules will be invaluable in sorting out what has occurred.

As schedule updates are prepared, they should be submitted with a brief narrative that describes the progress that has taken place during the update period, anticipated problems, changes in the sequence of the work, plans for recovering time if the schedule is behind, activities which are near critical, and so on. The narrative saves time by summarizing what a thorough review of the schedule update would show. It also demonstrates that the manager is aware of the progress and the problems and is taking positive steps to resolve the situation. An example of a schedule update narrative is shown in Exhibit 7-3.

NOVEMBER 28, 1992

SCHEDULE UPDATE

WATER POLLUTION CONTROL PLANT
PROJECT C-27-4102

PREPARED FOR

GENERAL CONTRACTOR

NOVEMBER 1992

I. *PROGRESS*

The following schedule update reflects job progress for the time period of October 20, 1992, through November 17, 1992. During this period, the excavation and backfilling, panel feeders, rough wire lighting, and the installation of fixtures were completed in Bay 4. In Bay 2, the demolition work has been completed and the backfilling and compaction is now under way. In addition, progress was made on the hollow metal doors and painting in Bay 4 and the rough wiring for future doors in Bay 5A.

The construction activities in which progress was made during the update period are shown in Table 1.

TABLE 1

ACT	DESCRIPTION	LOCA	OD	RD	FLOAT
General Contractor					
1220	Hollow Metal Doors	Bay 4	5	1	106
1230	Paint	Bay 4	15	8	99
2070	Demolish Bay 2	Bay 2	15	0	—
2080	Backfill & Compact	Bay 2	10	9	42
Electrical Contractor					
1182	Panel Feeders	Bay 4	3	0	—
1186	Rough Wiring Lighting	Bay 4	5	0	—
1188	Install Fixtures	Bay 4	3	0	—
1190	Rough Wire & Inst Recpt	Bay 4	3	2	97
3086	Rough Wire Future Doors	Bay 5A	5	3	85
OD	Original Duration				
RD	Remaining Duration				

Mechanical Contractor—No progress this period

The status of outstanding submittal items is shown in Table 2.

TABLE 2

ITEM	STATUS
General Contractor	
Wire Mesh Partitions	To Be Submitted
Metal Shelving Stor Cab	To Be Submitted
Mechanical Contractor	
Waste Oil Tank	Approval Pending
Paint & Spray Bth Fig. & Valves	Approval Pending

II. *CRITICAL ACTIVITIES*

This schedule update is indicating a projected project completion date of September 12, 1993—42 working days behind the contract completion date of July 15, 1993. The critical path runs through the relocation of the hydrant in Bay 3 and the completion of Bay 2 and Bay 5.

III. *LOGIC CHANGED*

The activities in Table 3 were agreed to the schedule logic to account for the issuance of change orders on the project.

TABLE 3

ACT	RESP	DESCRIPTION	LOC	DUR
2006	Gen	CO #1—Structural Repairs Predecessor—2000 (Conventional) Successor—2010 (Conventional)	Bay 1	35
2032	Gen	CO #1—Incr Slab to 8″ bet. Bays 1 & 2 Predecessor—2025 (Conventional) Successor—2030 (Finish-to-Finish)	Bay 1	5
2068	Gen	CO #2—Removal of Track & Additional Concrete Predecessor—2070 (Start-to-Start) Successor—2080 (Conventional)	Bay 2	3
1082	Gen	CO #2—Mods. to Grade Beam Predecessor—1080 (Conventional) Successor—1100 (Conventional)	Bay 4	28

IV. *ANTICIPATED WORK THIS MONTH*

During the next update period, the General Contractor will be working on the backfilling and compaction activities at Bay 2, along with the painting of the floors and walls at Bay 4. The mechanical contractor expects to be working on the 10″ water main to the street, as well as the turbine ventilators at Bay 4.

Exhibit 7-3

COST RECORDS

Obviously, all cost records are important. These include invoices, purchase orders, pay requests, quotations, payrolls, and change orders. There are many reasons why these should be maintained.

Should a project be terminated for convenience, the normal method of paying the contractor is for the costs incurred within the definitions of the contract. Effectively, a fixed price contract may be converted to a cost type of contract. If the contractor does not have records to support all costs, it may not be able to recover costs it is due.

If change orders are not resolved prior to the work being performed, the owner and the contractor may have to resolve them based on the cost of the work. Therefore, both parties should maintain adequate records for this.

Payroll records may be helpful if questions arise concerning the amount of labor expended during a project. In one situation where the design professional was sued for negligence, the contractor alleged a significant cost overrun in labor. The certified payrolls, along with the contractor's original estimate, were useful in resolving the issue.

PROJECT PROCEDURES MANUALS

This chapter began by noting that the manager should contemplate and plan to maintain information on the project at its inception. An effective method both for planning information management and for implementing it is the use of a project procedures manual. The manual lays out exactly the procedures to be followed on the project.

There have been books written on project manuals and sample approaches are also available. Unfortunately, a procedures manual is project specific. You cannot just use an off-the-shelf manual effectively to help manage the project. The project manager should be responsible for developing the manual.

Many different areas can be addressed in the manual. Exactly what is incorporated depends on the project, the user, the purpose of the manual, and the manager. A sample outline for a procedures manual is shown in Exhibit 7-4.

ACE CONSTRUCTION COMPANY

Project Procedure Manual
for the
Mississippi River Bridge
Contract C7354

I. Contract Considerations

  A. Changes and Extra Work

B. Progress Schedules
C. Quality Control/Quality Assurance
D. Submittals
E. Site Visits

II. Staffing and Authority

A. Project Manager
B. Project Superintendent
C. Owner's Representative

III. Project Recordkeeping

A. Daily Reports
B. Cost Reports
C. Correspondence
D. Memoranda
E. Progress/Status Reports
F. Speed Memo
G. Telephone Logs
H. Photographs

IV. Cost Monitoring and Control

A. Project Budget
B. Daily Cost Tracking
C. Cost Variances

V. Job Relations

A. Interaction with the Owner's Representative
B. Subcontractors
C. Suppliers
D. Outside Agencies

VI. Project Schedule

A. Format
B. Updates
C. Tracking Progress
D. Variances

Exhibit 7-4

To further illustrate the type of information contained in the manual, a portion of the section dealing with project documentation is shown in Exhibit 7-5.

III. Project Recordkeeping

A. Daily Reports

1. Daily reports will be filled out on the project by the Superintendent. The standard daily report form shown in Appendix A will be used. The purpose of the daily report is to record exactly what work took place, the size of all crews by trade, including subcontractors, the equipment on site, etc. The daily report form should be self-explanatory. Appendix A, however, includes examples of completed daily reports filled out both correctly and incorrectly.
2. The Project manager will review all daily reports for accuracy and completeness. Forms will be sent via FAX every day to the home office.
3. All work recorded on the daily report will be identified by the corresponding CPM schedule activity number. Progress should also be noted in units of work, such as cubic yards of concrete placed, tons of reinforcing steel, etc.
4. Particular attention should be paid to recording any problems or extra work items that may occur.

Exhibit 7-5

As can be seen in Exhibit 7-5, the information is clear and simple. The manual references sample forms included in it. It should be noted that this manual gives examples of both proper and improper forms. The idea is to show the staff what they should avoid besides giving an example of what is desired. The manual also specifically defines who is responsible for the forms and how and by whom they will be checked.

A procedures manual should be a working document. That is, it should be clear, simple, and helpful to the staff who will be implementing it. The goal is not to create a 300-page manual that no one can understand. Anyone who has read government publications will understand. Exhibit 7-6 shows an excerpt from an actual three-volume procedures manual developed for a project.

2. *CORRESPONDENCE FLOW, FILING, AND RETRIEVAL ACTIVITIES*

- Develop and implement a project-specific Correspondence Flow Matrix so that recordkeeping systems continue to operate smoothly during the project's Design Development Phase.
- Distribute copies of the project-specific Correspondence Flow and Filing Matrix to all project staff; the Owner's project Administrator, The

manager of Program- and Non-Program-Related Engineer and the Manager of Construction Engineering, Construction Management Services Group Manager; and the Owner's Project Administrator and the Design Consultant.

- Distribute copies of the project-specific Correspondence Flow Matrix to the Corps of Engineers, the Environmental Protection Agency, the Wisconsin Department of Natural Resources.
- Use the filing system set up for design data for adequate storage and retrieval of all design information and correspondence.
- Continually maintain the correspondence related to the project. File it during the Design Development Phase, and coordinate the retrieval of this information to be sure that the filing system conforms to the requirements of the Construction Management policies and procedures for filing and information retrieval contained in this manual.

RESPONSIBILITIES

- The Project Manager will use Responsibilities Checklist III-3-4, "Correspondence Flow" and Summary Checklist III-3-5, "Filing Considerations" in Part III, Section 6 of this manual to define standard tasks to be accomplished by project filing systems. This checklist will serve as a basis for the project-specific Correspondence Flow and Filing Matrix.
- The Project manager will prepare and implement a project-specific Correspondence Flow and Filing Matrix, and review it with the Project Design Manager, the design Services Coordinator, and the Owner's Project Administrator.
- The Project Manager will distribute copies of the Correspondence Flow and Filing matrix to all appropriate Program Management Office Staff; the Design Services Coordinator, the Project Design manager; the Owner's Project Administrator and Manager of the Program- and Non-Program-Related Engineering; and the Construction Management Services Group manager.
- The Project manager will continually monitor correspondence flow, filing of design correspondence, and retrieval of information to be sure it conforms to the Construction management policies and procedure requirements contained in this manual and the Program management Office standards.

Exhibit 7-6

The manual excerpted in Exhibit 7-6 is of questionable benefit. It lacks clarity, is rampant with bureaucratic terminology, and defines tasks for the project manager that probably should be delegated to others.

Make your procedures manual simple. Make it concise. Once it exceeds 100 pages, the odds are that no one will read it. If developed with care, and implemented with a positive and constructive attitude, the procedures manual can be a very effective tool to assist in the management of the project.

SUMMARY

Effective information management can be achieved only by careful up-front planning and diligent execution and monitoring throughout the project. In today's fast-paced highly competitive construction project, clear communications are an absolute necessity. The consequences of misunderstandings and misinterpretations are not worth the risks they create. The prudent manager will plan, structure, and implement a comprehensive documentation program at the outset of a project and refine, streamline, and supplement it throughout the execution of the job.

CHAPTER 8

STRATEGIES FOR PROBLEM SITUATIONS

No one book can address all the potential problems that a manager might encounter during the execution of a construction project. While many problem areas have been discussed earlier, this chapter addresses the most common and troublesome situations that can arise and offers some ideas on how the manager might handle these.

Problem situations are inherent in any construction project. The specific areas that will be addressed in this chapter all relate to the goals of completing the project on time, within budget, and in accordance with the plans and specifications. If any of these problems occur, the risk of not achieving one or more of these three goals is significantly increased.

TERMINATION AND DEFAULT

This subject was discussed briefly in Chapter 4. The reality of a termination situation, however, is extremely traumatic. Until you have experienced the full import of the situation, you can never completely understand it. There are two reasons why an owner would terminate a contractor or a contractor terminate a subcontractor: convenience or default.

A convenience termination can have many causes. These could include the decision to make major revisions in the overall scope of the contract and consequently needing to rebid the work, a realization that the continuation of the project is not cost effective, or the need to halt the project for environmental concerns. Presuming the existence of a termination for convenience clause in the contract, the exercise should not be particularly difficult as long as the contract provisions are followed. The important consideration in a termination for convenience is that the contract is effectively being changed from a lump-sum contract to a negotiated procurement or a cost type of contract. The contractor is paid based on the costs it has expended for the work and a reasonable profit and overhead. Normally, the most difficult part of a termination for convenience is the determination of the costs due. Almost invariably this requires an audit of the contractor's records and perhaps the same for subcontractors and suppliers.

A termination for default is an extremely serious decision and must be carefully reviewed before it is effected. In this course of action, you are determining that the contractor or subcontractor has materially breached the contract and, hence, is in default.

In order to determine if the contractor is in default, the manager must review the contract to see exactly what is specified as a reason for termination. Many contracts are very general in the wording of what constitutes default. While this gives the manager a great deal of leeway, it likewise gives the terminated party and its attorneys ample ammunition to contest the termination as wrongful. Termination for default should only be used as a last resort. Any other possible options should be attempted before termination is pursued. In this vein, a couple of examples may be beneficial.

On a project involving the construction of a complex semiconductor facility, the electrical prime contractor believed that significant changes had occurred on the job. The prime contract was of a performance nature and consequently any changes would have to be viewed within the context of the scope of work originally defined.

The electrical prime was aware of the importance of the rapid completion of the project and in order to motivate the owner to agree that the contract should be amended from a lump sum to a cost plus, the prime began a job slowdown. It reduced its on site construction personnel from 140 to 60. Naturally, this got the owner's attention.

The foremost consideration of the owner was to terminate the prime and bring on another to complete the work. Unfortunately, in most cases, this course of action takes some time and may even further delay the completion. It almost always increases the final cost to the owner. While the extra cost may be recoverable from the defaulting prime, the outcome will await a lengthy court process.

As an alternative to terminating the prime, the owner first sent a letter stating that it believed that the prime's actions of reducing the work force were unjustified, that there were contract mechanisms that allowed the prime to pursue increased costs or changes without slowing down or stopping the work, and that if the situation were not corrected within a specified period of time, the owner would consider pursuing the termination alternative. The same day the letter was delivered to the prime, the owner solicited the assistance of another prime electrical contractor who had been the second low bidder on the job originally. The owner agreed to pay the second prime for the assistance of a project manager, superintendent, and estimator to walk the job site and prepare an estimate of cost and time to finish the job. Those individuals performed their job site review and began preparing the proposal.

The original prime immediately learned the identities of these individuals and understood that the owner was serious and was considering enlisting the help of another contractor. As a consequence, the original prime requested a meeting with the owner and agreed to increase the staffing of the project in accordance with the schedule. In return, the owner agreed to consider a proposal from the prime for alleged changes in the scope of the work.

By enlisting the second prime, the owner accomplished two important tasks. First, it made clear to the original prime that it was serious and was proceeding with work necessary to terminate. Second, it allowed the owner to gain some time by getting a jump on what would have to be done anyway if a termination was effected. Furthermore, by having an estimate prepared for the completion of the work, the owner then had a measure of what the potential increased costs might be if a termination was effected. This allowed the owner to make a reasoned decision on the termination and also on how much it might be willing to consider of the cost proposal from the original prime. The bottom line was that the project was completed on time. The owner did grant some additional cost to the original prime which the prime was able to support as valid changes from the original scope. While the owner may have paid a few more dollars than originally envisioned, the prime contractor received a lot less than originally requested and most important, the project completion date was met.

The manager should always be thinking of alternative strategies to termination that will result in achieving the ultimate objectives at the least possible cost. Be creative.

In another termination situation, a general contractor was significantly behind schedule and had performed substandard work on a portion of the project that was already completed. The owner was keen to terminate the contractor and bring on a new general to complete the work and correct the deficient work.

Prior to effecting a termination, however, the owner requested a meeting with the general contractor. During the meeting the owner explained that it was extremely distressed at the late completion date projected for the project, the lack of effective scheduling, the substandard workmanship, and the absence of any recovery plan on the part of the contractor. The contractor's response was simply, "We'll fix it and finish on time except for the delays that you have caused." This did nothing to move the ball forward.

The owner enlisted the services of a new general contractor in much the same manner as in the previous example. Concurrently, the owner sent a letter to the general contractor informing it that the owner considered the contractor in default of the contract and requiring that the contractor meet with the owner and the owner's scheduling consultant to prepare a recovery schedule that would reflect how the remaining work would be completed and how the remedial work would be handled. The owner noted that the schedule to be developed would be the contractor's schedule and that the owner was only supplying the resources—the scheduler—to effect this.

The contractor initially agreed. On the morning of the meeting, however, the contractor did not attend, but instead sent a letter stating that it would not take part in the process and was going to complete the job via its own plan of action. It did not define what that plan of action was.

In response to this position, the contractor was terminated. A new general contractor was mobilized and working within a week. The project was completed late and at a significant cost overrun. The process ended up in a time-consuming and expensive arbitration from which the owner ultimately recovered $14,000,000 from the original contractor.

This second example obviously was not a successful use of alternative approaches. It does illustrate, however, that attempts are worth making, and that every effort should be made before effecting a termination. By at least going through the process, the contractor was given another chance and the owner was able to get a second general contractor on board early so that the work could continue almost uninterrupted.

The manager who considers effecting a termination for default must always reassess the situation and consider what alternatives can be pursued. This does not guarantee success in avoiding the termination.

It does increase the manager's chances of avoiding it, provides better preparation should it occur, and looks more reasonable if litigation ensues.

Many contractors consider if they should default on a project and just "walk away." In the vast majority of situations, this is not a prudent decision. A contractor validly can stop work on a project only if the owner or the other party to the contract is in default of the contract. For an owner to be in default it must either have failed to pay in accordance with the contract or have caused a cardinal change. If the contractor quits the project for any other reason, it will be in default of the contract. In the vast majority of cases, it is best for the contractor to continue work.

The reason that a contractor may decide to walk away from a project, absent a breach by the owner, is that it is not economically feasible for the contractor to continue. The contractor must recognize that in walking, it may be sued for default and may be liable for the reprocurement and completion costs in excess of the contract balance. If the contractor's financial position on the project is bad enough, stopping work might be the most cost-effective decision. By continuing to work, the contractor will be losing more money. By defaulting, the contractor may be sued, but most likely will be in bankruptcy already. The obvious kicker to this is if the project has a performance bond. In that case, the surety will most likely become involved. That involvement will vary greatly depending on the situation, the contractor, the amount of money at stake, and other factors. There are many different courses of action that a surety may take. This discussion will not attempt to cover all of them. A few, however, will be noted.

The involvement of the surety may depend to some extent on the actions of the owner. In some instances, the owner may terminate the contractor and take over completion of the project without allowing the surety to step in. An owner should seek legal counsel before heading down this path.

If the contractor has financial problems, the surety may provide financial assistance to insure project completion. Alternatively, if the contractor is terminated, the surety may choose to complete the contract work either with a new contractor or through the defaulted contractor. Overall, the contractor must fully understand the ramifications of its default on a contract. Effectively, default may put it out of business, either from a financial perspective or because surety companies might refuse to bond future work.

In any termination action it is important that both parties follow the provisions of the contract explicitly. Normally, the contract requires a notice of intent to terminate the contract. There is then a waiting period to provide the other side the opportunity to "cure" the breach. Only alter that can the termination be effected. If one party pursues a termination contrary to the provisions, it may later be viewed as an improper termination with all the ensuing legal consequences.

PROJECTS BEHIND SCHEDULE

Many projects fall behind the planned schedules for progress and completion. The reasons for this are many. Usually, a compounding factor is that the project was poorly scheduled and not accurately monitored during the course of construction. What does the manager do when the project is behind?

Obviously, the first consideration is the ramifications of a late completion. From the contractor's perspective, it may face liquidated damages. The manager must then assess if time can be gained on the schedule, how that can be done, what it will cost, and how that compares with the cost of being late. As noted in the discussion of cost-slope calculations in Chapter 5, a detailed evaluation of the schedule and costs may show that the cheaper alternative is to finish late. Financially, that is the best decision. The manager must consider some intangibles, however. For instance, late completion of a project may result in an adverse performance rating which could restrict bidding on future projects. It may also adversely affect the contractor's reputation in that particular area or industry or with that owner. This consideration may override the additional cost involved in accelerating the project.

Owners must also evaluate the ramifications of late project completion. All too often, a manager believes that a late finish is unacceptable regardless of the cost. It becomes a matter of pride. Forget pride; the question is simply one of economics, both long and short term. If it is more cost effective to finish late then this should be accepted. If the owner insists that the project must be finished by a certain date, however, then it must become an active participant in the process of assisting in achieving this goal. It is not enough just to tell the contractor to accelerate. The acceleration or recovery of time must be carefully planned and the owner should be aware of the plan and understand the feasibility of the plan. Whatever shape the original project schedule is in, a recovery schedule should be prepared that is meticulously detailed and constantly monitored for the duration of the job. It should be obvious to the manager that the project got to this point with some help from poor time management in the first place. To have any hope of recovering demands even better time management now.

In the process of planning to recover time, the manager must evaluate why the project is behind schedule. This, however, must not become the major focus of attention. While many potential reasons can be listed and evaluated, whether or not the exact reason is determined becomes something of an academic exercise since the main goal is to make up the time. Therefore, the manager might better concentrate on formulating alternative courses of action, evaluating those alternatives in terms of success and cost, and in making decisions to accomplish the final goal. Too many managers get bogged down in trying to determine at whom the finger should be pointed. Focus your efforts on tasks with positive benefits.

CLAIMS AND DISPUTES

Construction claims are simply unresolved change orders. If we had resolved them before they became a claim, the vehicle would have been a change order. Therefore, claims are handled the same way change orders are. Unfortunately, when an issue becomes a claim, the parties become less cooperative and less willing to resolve their differences. If at all possible, this should be avoided. To do this, the manager must exercise unusual self-discipline and avoid the temptation to draw a line in the sand and dare the other guy to step across. When a situation develops that the parties cannot agree on, it is common for one or both parties to internalize the problem as a personal one. This prompts the manager to communicate more aggressively and sometimes even threateningly with the other side. The expected response, in turn, may be even more aggressive.

Step back from the fray. The potential dispute is business, not personal. Use neutral language to avoid igniting the conflagration. For example, call the item in question an "unresolved change," a "request for an equitable adjustment," a "proposal for extra work," or another nonthreatening term. Don't stop communicating because that's when the attorneys will start talking. Try to understand the other party's perspective. That way you can best think creatively and develop an approach that you can sell the other side. Effectively, you want to devise a method that gives a "win-win" outcome to the potential claim situation.

For example, if a contractor believes that a differing site condition exists, it may draft a letter stating, "The extra work was required because of a grossly inadequate design. The Engineer clearly misrepresented the subsurface conditions." Obviously, this will alienate the engineer and will not facilitate resolution. As an alternative, the contractor could have drafted the letter to say, "The extra work involved some unusual subsurface conditions. No one would reasonably have anticipated them." This is a neutral approach.

Claims usually emanate when two parties disagree over whether a condition differed from that originally required by the contract. If the parties agree that the condition was different, or a change, then a claim may result from a lack of agreement over what the impacts of the change were or what the reasonable costs should be. All claims relate back to one or more of these three parts: Is it a change? What is the impact? What are the costs?

In any claim situation, all parties should document the issue as thoroughly and as contemporaneously as possible. Should the issue not be resolved and a formal dispute procedure be required, then this documentation will be invaluable.

Total cost or *total time claims* are another subject of concern. For a total cost claim, the contractor submits a claim for additional costs. In this claim, it asserts that the owner caused numerous problems and changes on the project. The contractor further asserts that it bid the job at a certain cost and because of the problems and changes the job actually cost more. It then claims that the owner owes the difference between the bid cost and the total cost.

This is not a strong argument for recovering additional cost. A total cost claim assumes that all problems are attributable to the other side and that none of the problems are attributable to you. Finally, it assumes that the bid or estimate was correct. Given the nature of these three assumptions, a total cost claim becomes difficult to support when attacked. In general, it should be avoided.

A total time claim is just a variation of a total cost claim. In this type of claim, the contractor again asserts all the problems that occurred and alleges that the owner is fully responsible. It then notes that it scheduled the job for a certain duration and that it took longer. The increase in duration is claimed to be owner-caused delay and, based on that, the costs are claimed. This is really the same thing as a total cost claim. It has the same problems and weaknesses and should also be avoided.

Contractors use the total time approach when they cannot readily recognize another approach to describe what occurred. Normally, the project did not have a good schedule, so there is no yardstick against which to measure delays with respect to specific causes or actions. This approach is also used when the contractor perceives that there were so many problems and changes that a specific delay presentation is impossible. Often, a total time claim is submitted because it is easy, fast, and inexpensive. Despite these reasons for its use, a total time claim should be avoided.

In general, the use of total time or total cost claims bespeaks a lack of good management during the project. If the manager had a good schedule it could have determined the delays and the reasons for them. If management kept good records and closely monitored the work, it could have determined what changes occurred and the specific impact of these changes. It must be noted that some exceptions may exist and in those cases the total cost approach may be applicable.

PUNCHLISTS AND RETAINAGE

All projects must pass some form of final inspection to insure that the work is completed in accordance with the contract documents. Normally, these inspections generate *punchlists*, lists of items to be corrected. The manager in charge of creating or overseeing the punchlist process should exercise care to insure that the process is carried out in an efficient, fair, and reasonable manner. In some instances, owners may generate series of punchlists and the process turns into a never-ending drill. Once the project is substantially complete and any other major items have been performed, a date should be scheduled jointly with the contractors to make the final punchlist inspection. The contractors should be present so that they understand what items the owner's representative considers deficiencies. In some cases, a list is created in the absence of the contractors and the comments then are unintelligible to them.

The punchlist should be a fair representation of those items that have not been performed in accordance with industry standards. Any items that are noted should be reasonable. The punchlist should not be a harassment of the contractor. Likewise, the process should not be one in search of perfection. No project is perfect.

If the owner has allowed other contractors to work in the areas or has had equipment or furniture moved into the project site, the punchlist should not require the contractor to correct problems that were caused by these outside sources. It is better to finish the punchlist before any other independent activities take place.

In one project during the punchlist phase, the owner's representative was particularly disenchanted with the painting inside the facility. As the parties walked the area, the owner's representative would circle areas—with a black marker—that he considered needed repainting. This required not only touchup paint but three coats to completely cover the black marker. Actions such as this can only be construed to be harassment, and should be avoided.

During the course of the project, the owner usually holds money as *retainage*. The exact amount of retainage is specified in the contract, and normally is in the range of 5% to 10% of the total cost. The purpose of retainage is to allow the owner to hold money for items that may not be completed by the contractor. Usually retainage is held for the punchlist items. The one note of caution in this is that the amount of the retainage should be representative of the cost of the punchlist work. For example, an owner may be holding $100,000 on a $2,000,000 project. When the punchlist is generated, the owner should be able to reasonably estimate the cost to perform this work should the contractor fail to do so. If a reasonable estimate is $15,000, there is no reason for the full $100,000 to be held. To be on the safe side, the owner could double the estimate and hold $30,000 but release the remaining $70,000 to the contractor. Many owners believe that retainage is leverage and use it just in that fashion. Effectively it is used as blackmail to have all items completed. This is not the concept of retainage and it should not be used in this manner.

BUDGETS

Most contractors establish budgets for the work on the project. Theoretically this budget is monitored against the actual cost expended so that the contractor is alerted to any variations and any possible problems. Unfortunately, the majority of the time, the budget versus actual tracking process is fraught with errors and problems. As a consequence, the contractor may not discover till it is too late that a significant overrun is going to occur or has already occurred. How can this happen?

Figure 8-1 illustrates the most common format for budgets and cost-tracking systems.

The project budget shown in Figure 8-1 has been broken down into several categories of work. Based on these categories and the original estimate, the contractor has budgeted specific amounts for the anticipated cost of performing the work. The next column records the ex-

PROJECT COST REPORT (Partial)							
COST CODE	WORK ITEM	BUDGET	CUM. COST LAST PERIOD	COST THIS PERIOD	COST TO DATE	% COMPLETE	VARIANCE
01000	MOBILIZATION	$12,000	$12,000	0	$12,000	100	0
02104	CLEAR & GRUB	$8,500	$7,800	0	$7,800	100	$700
02200	SITE GRADING	$6,000	$4,000	$2,400	$6,400	100	($400)
02254	EXCAVATE FOR FOOTINGS	$3,200	$2,500	$7,000	$9,500	100	($6,300)
03100	CONCRETE FOOTINGS	$28,000	$6,000	$17,600	$23,600	71	-
03200	CONCRETE GRADE BEAMS	$22,000	0	$14,000	$14,000	64	-
03300	SLAB ON GRADE	$38,000	0	0	0	0	-

Figure 8-1. Common cost-tracking system.

penditures during the last time period, usually a month. The succeeding columns record cost for this period, costs to date, the percentage completion of each category of work, and the variance. One problem with this system is that negative variance from the budget can only show up after it has been achieved. Also, while projections based on percentages versus expenditures can be made, they are often misleading or are readily rationalized if it appears that a negative variance is going to occur. For instance, if the projections indicate that costs are overrunning, the project manager may assert that costs are initially higher but should go down as the crews become more proficient in the performance of the tasks. Another problem that occurs is that the individual recording the information usually has an interest in the profitability of the project and does not want to look bad based on the monthly cost reports. As a result, if it appears that a negative variance might occur in a category of work, the recorder may shift some of the dollars to the succeeding category. For example, if an overrun is apparent in the foundations area, some dollars may be shaved from here and placed in the cost code for structural concrete. Ultimately, a huge overrun may appear in the project close out and clean up.

While there is no fail-safe system, a more prudent method of monitoring costs is to use productivity or unit measurements for the categories of work, as shown in Figure 8-2. Wherever possible, the items of work have been defined into total units and hours that were estimated. Based on this, a unit rate is established. The costs are still maintained, but the periodic reporting can now focus on the unit rate to date and the unit rate for the period. This is far more precise in predicting the chances for a negative variance and for contemporaneously measuring the performance. Because this system measures performance from a productivity perspective, it enables the manager to recognize early the potential for problems, to determine why productivity is less than estimated, and to make corrections that will result in cost savings for the remainder of that type of work.

While this system is more difficult to set up and more time consuming to monitor, it is worth the effort. Attempting to monitor costs or productivity based on the system shown in Figure 8-1 is courting disaster.

PROJECT EQUIPMENT

Owners and designers may enter a project wanting certain types of equipment to be used. For instance, the owner may have a particular brand of electric motors throughout its facility. Therefore, it would want to have the same brand used in the new addition to the facility that is to be constructed. The easiest solution would be to specify the brand name motors. This is more cost effective in terms of repair work and spare parts. The owner is concerned, however, that if it specifies a sole-source item such as the brand name motors and any problems occur in the availability or delivery of these, then the contractor might have a valid claim for delay to its work because of this. Also, the owner may be prevented from using a sole-source specification. This is particularly true of public agencies.

In order to overcome the problem but still get the brand name desired, the owner writes the specifications such that no other supplier's equipment can fulfill the requirements. This doesn't get the owner off the hook. By drafting a specification that restricts the motors to the one brand, the owner has an effective sole source and can still have the problems anticipated. To avoid this, it might be more realistic and less risky for the owner to procure the motors directly and supply them to the contractor doing the construction. By making the equipment "owner furnished," many potential problems can be avoided. The caution to the owner is that it must procure the equipment early enough so as not to cause a delay to the contractor's work.

PROJECT COST REPORT (Partial)

COST CODE	WORK ITEMS	UNIT	BUDGET QNTY.	BUDGET BUDGET	BUDGET UNIT COST	INSTALLED TO DATE QNTY.	INSTALLED TO DATE COST	INSTALLED TO DATE UNIT COST	INSTALLED THIS PERIOD QNTY.	INSTALLED THIS PERIOD COST	INSTALLED THIS PERIOD UNIT COST	% COMP.	PROJ. VAR BASED ON INSTLD. TO DATE
01000	MOBILIZATION	L.S.	1	$12,0000	$12,000.00	1	$12,000	$12,000.00	0	0	0	100	0
02104	CLEAR & GRUB	AC	2	$8,500	$4,250.00	2	$8,500	$4,250.00	0	0	0	100	0
02200	SITE GRADING	S.Y.	6,000	$6,000	$1.00	6000	$6,400	$1.07	2000	$2,140	$1.07	100	-400
02254	EXCAVATE FOR FOOTINGS	C.Y.	640	$3,200	$5.00	640	$3,200	$5.00	140	$700	$5.00	100	0
03100	CONCRETE FOOTINGS	C.Y.	70	$28,000	$400.00	50	$22,000	$440.00	40	$17,600	$440.00	71	-2,800
03200	CONCRETE GRADE BEAMS	C.Y.	44	$22,000	$500.00	28	$14,000	$500.00	28	$14,000	$500.00	64	0
03300	SLAB ON GRADE	S.F.	19,000	$38,000	$2.00	0	0	0	0	0	0	0	0

PROJECT COST REPORT (Partial)

COST CODE	WORK ITEMS	UNIT	BUDGET QNTY.	BUDGET BUDGET	BUDGET UNIT COST	INSTALLED TO DATE QNTY.	INSTALLED TO DATE COST	INSTALLED TO DATE UNIT COST	INSTALLED THIS PERIOD QNTY.	INSTALLED THIS PERIOD COST	INSTALLED THIS PERIOD UNIT COST	% COMP.	PROJ. VAR BASED ON INSTLD. TO DATE
01000	MOBILIZATION	LS	1	$12,000	$12,000	1	$12,000	$12,000	0	0	0	100	0
	MATERIAL	LS	1	$6,000	$6,000	1	$6,000	$6,000	0	0	0	100	0
	LABOR	LS	1	$4,000	$4,000	1	$4,000	$4,000	0	0	0	100	0
	EQUIPMENT	LS	1	$2,000	$2,000	1	$2,000	$2,000	0	0	0	100	0
02104	CLEAR & GRUB	AC	2	$8,500	$4,250	2	$8,500	$4,250	0	0	0	100	0
	MATERIAL	-	-	-	-	-	-	-	-	-	-	-	-
	LABOR	MH	100	$4,000	$40.00	100	$4,000	$40.00	0	0	0	100	0
	EQUIPMENT	HRS	100	$4,500	$45.00	100	$4,500	$45.00	0	0	0	100	0
02200	SITE GRADING	S.Y.	6,000	$6,000	$1.00	6,000	$6,400	$1.07	2,000	$2,140	$1.07	100	-400
	MATERIAL	-	-	-	-	-	-	-	-	-	-	-	-
	LABOR	MH	75	$3,000	$40.00	75	$3,200	$42.67	25	$1,070	$42.67	100	-200
	EQUIPMENT	HRS	75	$3,000	$40.00	75	$3,200	$42.67	25	$1,070	$42.67	100	-200
02254	EXCAVATE FOR FOOTINGS	C.Y.	640	$3,200	$5.00	640	$3,200	$5.00	140	$700	$5.00	100	0
	MATERIAL	-	-	-	-	-	-	-	-	-	-	-	-
	LABOR	MH	40	$1,600	$40.00	40	$1,600	$40.00	8.75	$350	$40.00	100	0
	EQUIPMENT	HRS	40	$1,600	$40.00	40	$1,600	$40.00	8.75	$350	$40.00	100	0
03200	CONCRETE GRADE BEAMS	C.Y.	44	$22,000	$500.00	28	$14,000	$500.00	28	$14,000	$500.00	64	0
	MATERIAL	C.Y.	44	$2,200	$50.00	28	$1,400	$50.00	28	$1,400	$50.00	64	0
	LABOR	C.Y.	44	$17,600	$400.00	28	$11,200	$400.00	28	$11,200	$400.00	64	0
	EQUIPMENT	C.Y.	44	$2,200	$50.00	28	$1,400	$50.00	28	$1,400	$50.00	64	0
03300	SLAB ON GRADE	S.F.	19,000	$38,000	$2.00	0	0	0	0	0	0	0	0
	MATERIAL	S.F.	19,000	$19,000	$1.00	0	0	0	0	0	0	0	0
	LABOR	S.F.	19,000	$17,100	.90	0	0	0	0	0	0	0	0
	EQUIPMENT	S.F.	19,000	1,900	.10	0	0	0	0	0	0	0	0

Figure 8-2. Amended budget.

This same approach is advisable for equipment that has a long lead time for manufacture and delivery. The designer should be able to identify these early and advise the owner that with the normal lead times, the project may not be able to be built in accordance with the schedule. For these items, the owner can initiate purchase orders and effect the procurement even before the construction contract is let for bid. To allow the contractor to plan its work, the bid documents should reflect which equipment the owner is supplying and the earliest date that the contractor can expect that equipment to be available. By building in a comfortable margin of safety in the time specified, the owner should preclude problems from arising.

One final concern about equipment should be mentioned. In sophisticated projects, some equipment must be specially fabricated. If the contractor is responsible for this, the owner should require that any purchase orders for this fabricated equipment include an "automatic assignment" clause. This means that if for any reason the owner wishes to take over the purchase order directly, it can do so by effecting the assignment clause of the purchase order. An example may illustrate how this could be helpful.

In a complex chemical process project, the owner and contractor became involved in a major dispute concerning the progress of the work and the performance of some items of proprietary equipment. The result was that the contractor was terminated for default. At that stage, the owner went to the specific vendors that were fabricating the equipment for the process plant. In several cases, the vendors refused to release the equipment to the owner since they had purchase orders only with the contractor. Had an assignment clause been incorporated into the purchase orders, this problem would have been prevented.

SUMMARY

Almost any area of a construction project can become a major problem if not managed correctly. The preceding discussion addressed those areas that might not be readily recognized as potential problems. It also was intended to encourage you to think in a broad perspective so that you anticipate potential problems and devise methods to prevent or resolve them. If managers do not anticipate these types of situations or if they do not approach them thoughtfully, the risk of significant problems increases. The key to reducing risk is to consider every possible area and to think creatively in the management of the project and problems.

CHAPTER 9

SUMMARY

This book has covered a considerable number of topics in relatively few pages. Hopefully, by this stage you have gleaned some useful ideas on how to improve management of a construction project. To summarize all the diverse areas that have been discussed, we must step back and reassess our role as the manager. The manager must think creatively and keep the "big picture" in view.

Many texts cover the basic management principles of planning, directing, staffing, controlling, organizing, and implementing. But this text has delved in those areas only with respect to the specifics of a construction project. It has purposefully avoided a generic discussion of management. The reason for this is simple. Putting aside all the MBA buzzwords, the manager has one vital role. The manager makes decisions. Every organization has individuals who plan, who staff, who control. But the key players in those organizations are the decision makers. They are the real managers.

As the manager of a construction project, you cannot be afraid to make decisions. Whether the decisions are right or wrong, the project will not move forward unless you exercise the fortitude to make decisions when they are necessary.

The discussions of the preceding eight chapters provide a framework to enable enlightened decisions. Your decisions should be tempered and enhanced by the thoughts that the narrative has considered. A manager of construction, like any other manager, is assisted in the decision-making process by knowing the ramifications of a certain decision and the risks associated with it. Figures 9-1, 9-2, 9-3, and 9-4 reproduce the risk triangles from Chapter 1.

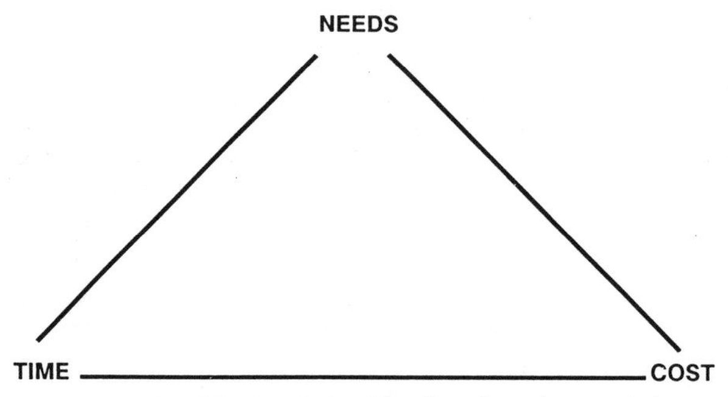

Figure 9-1. Goals triangle.

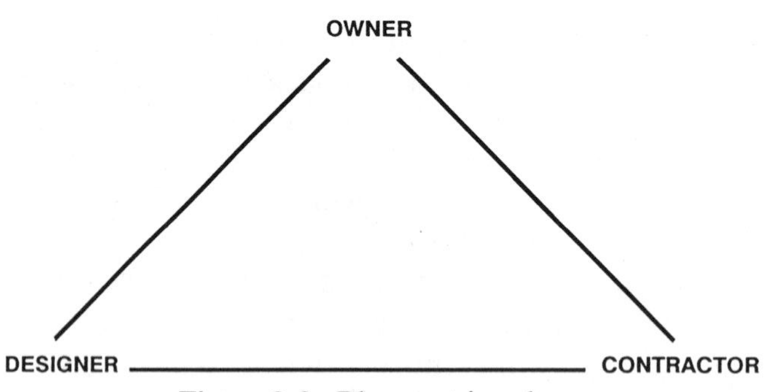

Figure 9-2. Players triangle.

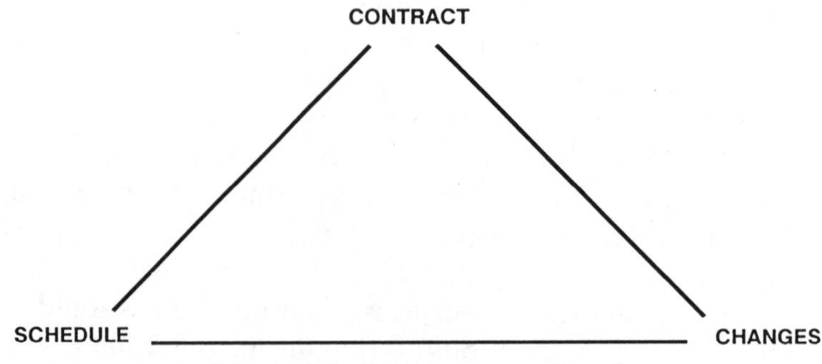

Figure 9-3. Project areas triangle.

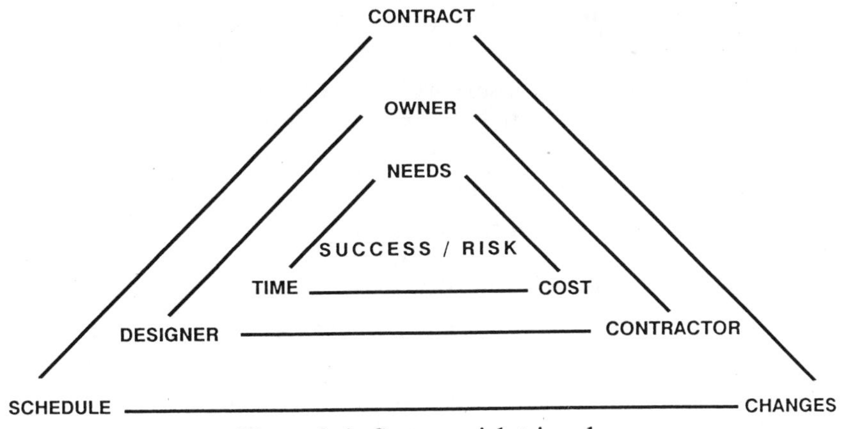

Figure 9-4. Success-risk triangle.

The success of the project is measured in the achievement of our goals with respect to needs, time, and cost. Virtually every construction project has three major parties: the owner, the contractor, and the designer. The respective managers for these three parties must have an understanding of the perspective of the other managers. Without this, their decisions may well create acrimony instead of accord. And to have a successful project, the three players must work together toward a common goal of getting the project built on time and within budget. A more subliminal but equally important goal is not to avoid ending up in court. If we allow our biases to drive our decisions, we will move away from that goal.

In the decision-making process, the manager must focus on the three major elements that require the most attention: the contract, the schedule, and the management of changes. If these three areas are handled in a professional and educated fashion, the chances of success are greatly increased.

For years, the author has presented seminars and training sessions on the topics in this book. At almost all of these programs, the attendees want to leave with a checklist of what to do to guarantee success. Unfortunately, no such checklist exists. If one did, there would be no need for managers because decisions would not require a leader to make them. Instead the decisions could be made from a flow chart.

The construction of a new project is a creative endeavor. It requires imagination and creativity with heavy doses of experience and reality. The managers of a construction project do not deal with the more esoteric areas that one sees in the research and development arena. They manage known processes, components, and systems. But despite the fact that the manager deals with knowns, the project cannot be managed by rote. The effective and successful manager of a construction project will take these known entities and put them together to yield a cost-effective project. One which fulfills the user's needs and in some cases one that is a unique facility. Construction project managers can never turn off their creativity. They cannot succumb to believing that their project is no different from any other that can be managed with a "cookie cutter" mentality. The manager should always view each project as a unique product. In that manner the manager will stay tuned to new ideas and new approaches and intertwine those with personal experience and the realities of the construction industry. The experience that must be applied is based on the pathologies of projects that went sour or did not survive. The realities must incorporate the existing available materials, equipment, and labor pools.

The most successful projects today are those where the manager approaches the job as if it had never been done before, avoiding stereotypes, rules of thumb, and artificial "dos and don'ts." The manager anticipates every potential problem and assesses potential solutions without biases. That manager of construction maintains an open mind, listens more than speaks, and makes decisions accordingly.

BIBLIOGRAPHY

Antill, James M. and Ronald W. Woodhead. 1990. *Critical Path Methods in Construction Practice.* 4th ed. New York, Canada: John Wiley and Sons, Inc.

Associated General Contractors of America. 1976. *The Use of CPM in Construction.* Washington, D.C.: Associated General Contracts of America.

Bent, James A. and Albert Thumann. 1989. *Project Management for Engineering and Construction*. Lilburn, Georgia: Fairmount Press, Inc.

Clough, Richard H. and Glenn A. Sears. 1991. *Construction Project Management.* 3d ed. New York, Canada: John Wiley and Sons, Inc.

Curie, Overton A. and Neal J. Sweeney. 1992. *1992 Wiley Construction Law Update.* New York: John Wiley and Sons, Inc.

DeGoff, Robert A. and Howard A. Friedman. 1985. *Construction Management—Basic Principles for Architects, Engineers, and Owners.* New York, Canada: John Wiley and Sons, Inc.

Frien, Joseph. 1980. *Handbook of Construction Management and Organization*. 2d ed. New York: Litton Educational Publishing, Inc.

Kerzner, Harold. 1989. *Project Management—A Systems Approach to Planning, Scheduling, and Controlling.* 3d ed. New York, Australia, England: Van Nostrand Reinhold Company, Inc.

Moder, Joseph J., Cecil R. Phillips, and Edward W. Davis. 1983. *Project Management with CPM, PERT, and Precedence Diagramming.* 3d ed. New York, Australia, England: Van Nostrand Reinhold Company, Inc.

Stephenson, James E., Esq., ed. 1990. *Alternate Clauses to Standard Construction Contracts.* New York, Canada: John Wiley and Sons, Inc.

Trauner, Theodore J., Jr. 1990. *Construction Delays.* Kingston, Mass.: R.S. Means Company, Inc.

Trauner, Theodore J., Jr. and Michael E. Payne, Esq. 1988. *Bidding and Managing Government Construction.* New York: R.S. Means Company, Inc.

Weist, Jerome D. and Ferdinand K. Levy. 1977. *A Management Guide to PERT and CPM.* 2d ed. Englewood Cliffs, NJ: Prentice Hall, Inc.

INDEX